Strong Family and Low Fertility: A Paradox?

European Studies of Population

VOLUME 14

The book series *European Studies of Population (ESPO)* aims at disseminating population and family research, with special relevance for Europe. It may analyse past, present and/or future trends, as well as their determinants and consequences.
The character of the series is multidisciplinary, including formal demographic analyses, as well as social, economic and/or historical population and family studies.
The following types of studies are of primary importance: (a) internationally relevant studies, (b) European comparative studies, (c) innovative theoretical and methodological studies, and (d) policy-relevant scientific studies. The series may include monographs, edited volumes and reference works.
The book series is published under the auspices of the *European Associaiton for Population Studies (EAPS)*

Editorial Offices:
Gijs *Beets*
Netherlands Interdisciplinary
Demographic Institute (NIDI)
PO Box 11650
NL-2502 AR The Hague, Netherlands
Phone.: +31 70 356 5200
Fax.: +31 70 364 7187
E-mail: beets@nidi.nl

Technical Coordinator:
Jacqueline *van der Helm*

The titles published in this series are listed at the end of this volume

Strong Family and Low Fertility: A Paradox?

New Perspectives in Interpreting Contemporary Family and Reproductive Behaviour

edited by

GIANPIERO DALLA ZUANNA
University of Padua,
Italy

and

GIUSEPPE A. MICHELI
Catholic University of Milan,
Italy

KLUWER ACADEMIC PUBLISHERS
DORDRECHT / BOSTON / LONDON

A C.I.P. Catalogue record for this book is available from the Library of Congress.

ISBN 978-90-481-6730-2 (PB)
ISBN 978-1-4020-2837-3 (e-book)

Published by Kluwer Academic Publishers,
P.O. Box 17, 3300 AA Dordrecht, The Netherlands.

Sold and distributed in North, Central and South America
by Kluwer Academic Publishers,
101 Philip Drive, Norwell, MA 02061, U.S.A.

In all other countries, sold and distributed
by Kluwer Academic Publishers,
P.O. Box 322, 3300 AH Dordrecht, The Netherlands.

Printed on acid-free paper

Printed in the Netherlands.

TABLE OF CONTENTS

[I] Revised, original by *Population and Development Review*, 1998, 24(2), 203-234 with the permission of the author.
[II] Reprinted *from* *Demographic Research*, 2000, 3(13), 1-27.
[III] Revised, original by *Demographic Research*, 2001, 4(5), 1-21.

GIANPIERO DALLA ZUANNA AND GIUSEPPE A. MICHELI

INTRODUCTION

New perspectives in interpreting contemporary family and reproductive behaviour of Mediterranean Europe

1. THE NEW GEOGRAPHY OF FERTILITY AND THE FAMILY IN EUROPE

The countries of southern Europe have begun to reduce conjugal fertility at a later date compared to most other nations in the west. This has been explained by means of the category of delay: the backwardness of the processes of accumulation and economic development being seen as the cause of the maintaining of the reproductive models of the past. Moreover, the influence of the Catholic Church in Italy, Spain and Portugal is supposed to have delayed the processes of secularisation, rendering difficult the changes in mentality necessary for assuming modern patterns of reproductive behaviour not only for fertility, but also for the variables which are strictly linked to it, such as sexuality, contraception and abortion (Livi Bacci, 1977; Lesthaeghe and Wilson, 1986).

1.1. The trends of very low fertility

Now the panorama is very different. Since the mid-seventies, southern Europe has been washed by the tide of a lowest-low fertility (i.e., TFR under 1.5 for several a prolonged period, Billari *et al.*, 2003), which in some areas has reached and maintained scarcely imaginable levels[1] for years on end. Conversely, other areas of Europe, where fertility started to fall many decades earlier than in the regions of the sourth, have recovered or maintained considerably higher levels of fertility, often close to replacement level. At the same time, there is little evidence of an inverse association between development and fertility. For example, in southern Italy, the economic system is still fragile and development is slow, and yet fertility in the last decade of the twentieth century is much lower in many of its provinces than in certain wealthy areas of France or northern Europe. The 'classical' interpretations

G. Dalla Zuanna and G.A. Micheli (eds.), Strong Family and Low Fertility: A Paradox?, 7-21.

Moreover, it is necessary to seek distinct explanations for the interpretation of the differences among countries and for that of the differences among couples within each country. Indeed, within each European country, the usual categories of interpretation maintain certain validity: fertility is lower among the more educated couples who are more consumer-minded and in which the woman is more involved in work and career.[2] If we compare the different countries, on the other hand, fertility is lower in the less educated nations and in those with lower percentages of working women (Pinnelli, 1992; Di Giulio *et al.*, 2000; De Rose and Racioppi, 2001).

In addition to this apparent paradox, there is another significant difference compared to the past. As we have seen, low fertility has traditionally been seen as one of the consequences of modernity, together with other changes in marital and reproductive behaviour. Fertility should therefore be lower where the constituent elements of the traditional family are weak. If this key of interpretation were still valid today, there would be a lower rate of fertility in the presence of higher rates of cohabitation, divorce and births outside marriage, i.e. the most striking phenomena of the second demographic transition (van de Kaa, 1987). In the final decades of the twentieth century in Europe, exactly the opposite took place: in countries where the changes were most widespread (such as those of northern Europe), a fertility close to replacement level coexisted with high rates of cohabitation, divorce and extramarital births, while in the countries of southern Europe, the very low fertility accompanied a situation in which 'new' forms of behaviour made slow and laborious headway (Cantisani and Dalla Zuanna, 1999). Once again, however, these regular features – marked when the comparison is among different European countries – are not automatically repeated in the case of couples. In short, what has happened over the past few decades has undermined models of interpretation of low fertility, which previously appeared well founded. In particular, it is necessary to explain the results that are obtained from the comparison of different European countries, with a very low rate of fertility in societies that are more anchored to more traditional modes of conjugal behaviour.

An interesting attempt to resolve this interpretative *impasse* has been made by McDonald (2000); in this study the emphasis is placed on the differences between public and private gender systems. According to this author, in societies of advanced development fertility may be at (relatively) high levels if the power relations between men and women are sufficiently balanced, within the family and the couple (the 'private' gender system) as well as in society (the 'public' gender system). In the societies of southern Europe (but also elsewhere, as in Japan and in other highly developed Asian countries), the barriers to equality between the sexes have apparently now fallen in the public sphere – with a consequent increase of women in places of responsibility and, more in general, in women's employment and expectations concerning the world outside the family – while inequality persists in the private sphere, with a gender system strongly biased in favour of the man. This attempt to reconcile macro and micro levels is extremely interesting, even though we are of the opinion that it requires more substantial empirical testing. Comparative studies using FFS data from various European countries have also shown that discrepancies may be observed between micro and macro results (De Rose and Racioppi, 2001). In those countries where the gender system (whether public or private) is more balanced, fertility is higher. However, within those countries, it is the more 'unbalanced' couples, in the traditional sense (in which the woman is a housewife, the man has little involvement in domestic tasks and opinions about life and the family are more traditional) that have more children.[3]

Moreover, it has yet to be explained how different countries with similar levels of economic development have been able to develop such differentiated gender systems. Europe appears to be experiencing a phenomenon which has been observed in India: in the Punjab, India's wealthiest state, economic development has not led to a reduction in forms of gender discrimination; on the contrary, in certain cases it has accentuated them (Das Gupta, 1987). There is strong evidence for the existence of basic anthropological structures, deeply rooted in the past, which regulate intimate interpersonal relationships and somehow manage to influence processes of modernization, accelerating them or slowing them down. Moreover, while modernization may indeed influence these basic structures, this does not necessarily lead to the ironing out of differences among countries, as a somewhat naive interpretation of the process of development might suggest.

Interpretative problems also arise when considering other possible explanations for lowest-low fertility and marital behaviour in southern European countries. The peculiarities of these countries have been attributed by some authors to the economic evolution, specifically to the high rates of unemployment and the difficulty of acquiring affordable housing. The poor access to the labour market and a shortage of affordable rental housing would compel many young people to remain in their parents' family for a long time (Billari *et al.*, 2003). These interpretations – certainly interesting and noteworthy – begin to falter when the available data are analysed in a more detailed way. In large areas of Italy (eg, all the regions of the north) unemployment among the youth has been virtually non-existent for the last twenty years. The high unemployment rates relating to Italian youth are skewed by the very high number of unemployed people living in the regions of the south of Italy. Yet, Italian youth in the south leave the parental home earlier than youth in the north. Besides, although in Italy and in Spain there is little alternative to buying a house, parents very often offer their children substantial financial assistance with the purchase of a new house (Barbagli *et al.*, 2003). Nevertheless, a considerable percentage of Italian youth (in the south and in the north) spend many years living with their parents while employed in a steady and well-paid job, despite having the option of acquiring a house.

Other possible interpretations of the peculiarities of these Mediterranean countries refer to the organisation of the welfare system. Some authors have highlighted how the geography of European fertility coincides with that of the diverse welfare systems (Esping-Andersen, 1990, 1999). The countries which are characterised by a family-oriented welfare state (mainly Spain and Italy) are also those that showed a very low birth rate in the last decades of the twentieth century. These countries do not reflect so much a low social expenditure as a favouring for social expenditure on the elderly (mainly pensions and health services) rather than on the youth (unemployment benefits, financial support at leaving the parental home, students' grants, facilitation for home building). Moreover, in these countries family benefits and – more generally – public transfers in favour of couples with more children are uncommon. These diverse welfare systems would appear to heavily condition reproductive and conjugal choices, tending to delay the exit of youth from the parental family and raise the living costs of having one more children.

This compelling interpretation also lends itself to criticism. Comparative studies show how difficult it is to establish a direct connection between fertility and public support in favour of families with more children, at least in western countries (Gauthier, 2002). Moreover, the reproductive behaviour of people born in Mediterranean Europe is remarkable even when they relocate to other countries. A study on the second generation migrants in Australia has highlighted that – welfare system and external conditions (housing market, labour market, generally residing in metropolitan areas of Melbourne and Sidney) being equal – the reproductive and conjugal behaviour of the children of the Italians and the Greeks differs considerably in comparison to that of the children of the immigrants from central and northern Europe, displaying a much lower proportion of cohabitations with partners or friends, or of people living alone, and lower fertility (Table 1). The particularly low fertility of immigrants of Italian origin has been recorded also in other contexts, such as Belgium (Perrin and Poulain, 2002, p. 44).

A more detailed analysis of the social system would therefore need to address the classical chicken and egg debate: are people – in some way – compelled to strengthen their family ties in order to adapt themselves to a welfare system they are certainly not capable of changing, or has the welfare system adjusted to a society based on strong family ties?

Table 1. Some demographic behaviours of people aged 25-29, born and living in Australia, by country of their parents (born in the same country) and gender. Census data of 1996.

	Australia	UK	Ireland	New Zealand	Netherlands	Germany	Hungary	Poland	Greece	Italy
					Never married					
M	62	63	69	73	51	65	69	70	62	50
W	46	46	51	60	39	50	53	52	43	39
					Cohabitation (among people living with his/her partner)					
M	29	32	33	40	20	27	25	25	3	6
W	22	25	25	34	18	24	22	17	1	2
					Alone or with friend (among people not living with a partner)					
M	56	47	53	56	45	41	36	29	12	16
W	49	50	45	61	53	38	39	39	12	18
					Mean number of children ever born					
W	0.86	0.84	0.63	0.57	0.90	0.64	0.62	0.45	0.44	0.46

Source: Khoo *et al.*, 2002.

In other words, why is it that over the last decades a family-oriented welfare system, a housing market based on property houses, and almost nonexistent state support towards families with more children have become part of the distinctive aspects of the countries in Mediterranean Europe? Why is it that in these countries the support and the financial help of the parental family heavily condition the access to adulthood and the building of human capital?

1.2. The geography of the strong family

The above observations will be useful in explaining the interpretative approaches suggested in the following pages. In our opinion, in order to understand fertility in Europe over the past 30 years, it is necessary to focus on the differences between European countries in the rules determining family relationships. Italy and the Iberian Peninsula are – now as in the past – the domain of the 'strong family', displaying marked temporal continuity in ties between parents and children. In contrast, in the other European countries, the ties between parents and children are already considerably loosened during the course of adolescence.

This difference goes back a long way, with its roots in the ancient differences between the Latin family and the Germanic family. It contributes to determining the differences in social organisation and mentality between the populations of northern and southern Europe. In the contribution of Reher and in the first contribution of Micheli, this subject is dealt with thoroughly, with an emphasis on historical and anthropological perspectives. These two authors show how the cut-off point between 'strong' and 'weak' families runs along the peaks of the Alps and Pyrenees, and is transversal as regards the different family forms and any other differences between the countries of Europe in terms of social organisation. For example, the boundaries between the two systems do not coincide with the split between Catholic and Protestant religions, even though Luther's incendiary ideas found a more sympathetic reception in the weak family structure. Moreover, the strong family persists down the centuries in Italy and in the Iberian Peninsula, independently of the rules governing residence after marriage (neolocal or patrilocal) and of family type (extended or nuclear), albeit with the regional features highlighted by the two authors.

Table 2. Frequency with which the elderly (65+) see family members in a few European countries.

	Every day	Twice a week or more
Italy	70.7	14.4
Greece	64.8	9.7
Spain	60.7	15.5
Portugal	59.8	9.6
Germany	46.5	15.6
France	34.2	16.2
Netherlands	19.2	26.6
Denmark	13.8	26.0

Source: Commission of the European Communities, 1993.

Above all, the diffusion throughout Europe of the rule of neolocal residence was not accompanied by the spread of the weak family. In other words, the strong family has not been weakened by modernisation. On the contrary, with emancipation from need and the arrival of prosperity, the system of the strong family was able to consolidate. Strong ties between parents and children have often helped accelerate the processes of development, thanks to economic forms of solidarity whereby the family maintains (and sometimes accentuates) its characteristics as an economic concern (or 'firm'). Moreover, the strong family may become an important instrument of social protection for all its members, helping prevent the isolation of the elderly. As we may observe in the example of Table 2, the split between northern and southern Europe as regards these variables could not be clearer.[4]

The strong family is not only a form of organisation of the system of family relations: it also makes a significant contribution to the construction of mentalities and social norms, which regulate the coexistence of persons in the countries of southern Europe. The strong-family areas are characterised by a system of values and social norms prioritising the good of the family as opposed to the good of the individual and that of other social aggregates (and of society *tout court*). Let us dwell for a moment on this aspect: as we shall see, it is important for our understanding of the different interpretations of the links between strong family and low fertility.

1.3. The two dimensions of the category of familism

It is now over 40 years since Banfield (1958) coined the category of amoral familism, in order to understand the pattern of the collective behaviour of a fragment of Mediterranean Europe. A quarter of a century on from the lively debate on the scientific validity of the thesis of this American anthropologist, his key of interpretation has turned out to be much more robust than deemed by his detractors. It is certainly weak if regarded as absolute and independent of historical context. But to reject it in the name of its presumed incompatibility with history as seen from the viewpoint of social actors, as other authoritative scholars of the Italian south have done[5], has been, in our opinion, a mistaken strategy, lacking in foresight, as it is the very ethos of familism that historians of social actors and practices are now having to take into account, in order to understand the specific nature of the model of development of these lands.

And yet the suspicion is legitimate that so frequent a use today of the category of familism leads to a loss of awareness of the double conceptual dimension implicit in the term as coined in 1958. Indeed, Banfield wrote (1958):

> The hypothesis is that the Montegranesi behave as though they were observing the following rule: maximise the short term material assets of the nuclear family, acting on the assumption that all others behave as you do. We shall call those behaving like this 'moral familists'.

And a little further on:

> We cannot easily say that an adult has his own individuality without bearing in mind the family to which he belongs. He exists not as Ego, but as parent. (...) Friends are a luxury that the Montegranesi cannot afford. (...) Friends and neighbours are not only potentially expensive, but also dangerous.

The category of familism therefore rests on two quite distinct dimensions, corresponding to the two rules of behaviour defined by Banfield. On the one hand, the good of the family comes before that of the collectivity, and on the other the good of the family comes before (and coincides, at least formally, with) that of its individual members, ego included. On the one hand we have amoral familism, in opposition to the Tocquevillian civic sense, and on the other, the philosophy of the stem-family, solidarity and intergenerational and blood pacts. Banfield put his finger on the first of the two aspects. Reher (see contribution in this volume), noting the family's extraordinary resources of self-organisation, has highlighted the persistence of the second component.

Figure 1. The two holons of altruism expansion in European models of social organisation.[6]

Subordination→ to community To the family ↓	No	Yes
No	(A) Individualism, philosophy of the 'particular'	(B) Tocquevillian communitarianism (Atlantic model)
Yes	(C) Amoral familism (Latin/Mediterranean models)	---

We may better note the similarities and differences between the Mediterranean model of familism and the north European/Atlantic one of Tocquevillian communitarianism[7] by highlighting the two different 'holons'[8], the two different forms of social organisation situated at a higher level than the individual on the hierarchical ladder, in which the individual himself primarily invests his own relational resources: the family and the community (Figure 1).

In short, Banfield's familism is not to be confused with an individually selfish pattern of behaviour (it is the family holon which follows this pattern): on the contrary, it is – as Banfield himself stressed – a variant of altruism, in which the social circle of reference is the circle of strong and blood ties.

2. FROM THE STRONG FAMILY TO LOW FERTILITY

If we observe the European geography of the strong family (with its normative systems generated by the familist logic of behaviour) and that of low fertility at the end of the twentieth century, the coincidences are surprising. However, it is no simple task to clarify the behavioural processes underlying this geographical correspondence. This volume contains two different possible interpretations, which, though departing from similar premises, reach quite distinct conclusions.

The first interpretation (see the second contribution of Micheli) supplements the anthropological continuity of the strong-tie models with an alternative way of interpreting the change, whereby the normal process of rational decision-making is superimposed by mechanisms which are "completely or largely based on normative/affective considerations, not merely with regard to selection of goals, but also of means" (Etzioni, 1988). The second interpretation (see the contribution of Dalla Zuanna), attributes more importance to the category of continuity: the strong family and familism have not changed very much over the past few decades in southern Europe, but the processes of modernisation have increased the value of a child to a much greater extent than in central/northern Europe.

A complete reading of Dalla Zuanna's contribution and of Micheli's second contribution will assist the reader in assessing the plausibility and empirical solidity of the interpretations put forward here. Over the next paragraphs we shall summarise the two schemes. Starting off from these two different interpretations, we shall also seek to answer a few questions on the future evolution of fertility in the lands of the strong family.

2.1. The Mediterranean model, between familism and particularism

In order to understand why the explanation of the transformations underway in Mediterranean society is to be sought amid many different patterns of action, it is useful to distinguish between the two dimensions, noted previously, of the anthropological category of familism – at least analytically. Indeed, the healthy state of the family unit is plain for all to see, as is its central role in organising the processes of social and economic reproduction, in those Mediterranean societies where community-orientation and 'civicness' are subordinated to the 'good of the family'. And the fact that the family is once again at the centre, at the very time when society is appearing to break finally away from the stable, robust model of the stem-family, is food for thought. While over the long term this model – and the virtuous familism which represents its philosophy of action – have contributed to the development of a demographically smaller family,[9] the cause-effect relation seems to have changed direction lately, and the extreme nuclearization of the family is becoming the springboard for the relaunching of the stem-family (see Micheli, first contribution), but in a variant based on a philosophy of the 'particular'.

We may attempt to explain this variant by making use of a metaphor drawn from economics. The differences between the broad areas into which Europe is divided in terms of family models (discussed in Micheli, first contribution) should be thought of as the result of the aggregation of individual choices, combining different types of factors capable of producing social organisation: those who will invest more in maintaining blood ties within the family cell, even to the (possible) detriment of collective interests (cell C of Figure 1), and those, on the other hand, who will invest more in the functioning of the community, even to the (possible) detriment of the pact of solidarity between the generations (cell B of Figure 1).

With the substitution of individual priority for familistic priority (from cell C to A in Figure 1), the Mediterranean model perhaps undergoes a radical change: this would not longer be a case of 'amoral familism', but rather of 'amoral individualism', as Banfield correctly pointed out (1958).[10] The difference between the Atlantic model and the Mediterranean model, which used to manifest itself primarily in a difference of 'technology of social organisation' (from more community-intensive to more family-intensive), now also features a different intensity of overall investment in 'altruistic relationality'.[11] Nor is this an unpredictable variant of the Mediterranean model: the idea of an amoral individualism, with priority attributed to individual interest rather than that of the community, was already in existence five centuries ago in the concept of the 'particular' ("particulare") utilised by Guicciardini[12] in order to synthesise the behaviour pattern of the Italian political class of his time.[13]

The new family model resulting from the shift from 'team' familism to an amoral individual familism would now appear to constitute an extraordinarily stable model, precisely because it is rationally optimum. But how may we explain the drift over the past few years, which has led inexorably to this new fixed endpoint? The slide towards forms of individualism goes well beyond the borders of Italy or southern Europe, and concerns the whole of Western civilisation, to a certain extent, as it has undergone the great ideational change of modernisation. So what is it that explains the peculiar nature of Italian society?

The second contribution by Micheli advances the hypothesis that in order to understand the shifts, which have marked Italian demography at the end of the twentieth century, it is necessary to descend beneath the surface-crust of the processes of rational decision-making and dig down into our deep psychology. Indeed, there are increasing signs of a growing hesitation (or fear) among the younger generations when they are faced with irreversible choices. Ginsborg (1998) maintains "very often, it has not been the lack of a sense of responsibility, but on the contrary an excess of it which has been the main contribution towards the decline of the rate of fertility". What meaning should we attribute to this increased sense of responsibility? Is it really a sign of greater control over one's own choices, or is it perhaps sometimes the sign of a loss of the ability to 'let go', to expose one's own actions to the risk of consequences which cannot be entirely controlled?

The second contribution of Micheli certainly doesn't deny the existence of a broad range of structural factors – economic, sociological and anthropological –, which is decisive in stimulating the rational assumption of 'parsimonious' choices. Rather, it questions the self-sufficiency of the dominant psycho-economic keys of interpretation. In particular, it is suggested that the crucial demographic choices of a life course are often 'decisions not to choose', in which there is no direct link between result and intention, and the completion of the transition from preferences to action requires a relaxation of the control of reason. The rarefaction of procreative behaviour may then be read as the result of a mechanism of interception between the preference system and the decisions, so that the individual appears lacking in reactivity, with no reasonable motives. The growing tendency to remain inside the niche of a pre-adult state would therefore find an added (though not exclusive) explanation in a kind of blocking of dispositional states: something like the letting out of the clutch, which makes it impossible to make a vehicle move forward, however desperately we put our foot down on the accelerator.

Bowlby's theory of attachment (1968) provides us with a possible translation of the concept of 'interceptor': if a cohort's experience of attachment has been in some way disturbed in the early years of life, it may actually produce a change (anxious insecurity) in this cohort or a partial de-activation (avoidant insecurity) of the instinct of attachment and also of its interface, the caring instinct. And an anxious or avoiding assumption of the caring instinct may also explain the recent changes in transition to the adult state.

2.2. The accentuation of familism

As we have already said, the interpretation of lowest-low fertility proposed in these pages by Dalla Zuanna gives priority to the category of continuity. Over the last 30-40 years, the same social processes have affected the whole of the west: reinforcement of extra-domestic roles of the woman, increase in prosperity and consumption, emphasis on post-materialist values. These processes have weakened the social significance of the conjugal bond everywhere. However, in the lands of the strong family, these processes have come up against the familist social structure, which has managed – at least in part – to slow them down and modify their effect in terms of demographic behaviour. For example, in weak-family regions, the loss of meaning of the institution of marriage has translated into an increase in cohabitation, while in Italy and on the Iberian Peninsula, on the other hand, children remain in their parents' family even beyond the third decade of life.

In this perspective, low fertility could be the fruit of a double process. On the one hand, as we have already said, the slower entry into adulthood is thought to have led to the putting off of the beginning of the reproductive career. On the other hand, the well-known trade-off between quantity and quality of children is particularly accentuated in Italy. Familist parents are not prepared to have children whom they regard as of 'low quality', and as a result they prefer to have one or (maximum) two, without depriving them of anything. The 'value' of children would therefore appear to have increased even faster than economic prosperity, precisely because parents continue to see in their children a prolongation of themselves. This reading of low fertility is not inconsistent with the more economic interpretations (such as those proposed by Becker or Easterlin). The very low fertility of southern Europe at the end of the twentieth century is thought to be the product of couples' desire to reconcile the co-existence of economic rationality with familist rationality.

Moreover, other processes deriving from the strong family system could induce in couples an accentuated neo-Malthusian prudence. For example, if parents regard children as their property and a prolongation of themselves, it is difficult to conceive of fiscal wealth redistribution on behalf of those families with more children. However, this kind of state intervention is easier to propose and justify if children are (also) recognised as being of social and collective value. It is perhaps no coincidence that strong-family countries are precisely those in which families with a greater number of dependent children are also those most heavily penalised by the taxation system.

So the processes of post-modernisation, which induce low fertility, are not thought to be substantially different in those regions of Europe with strong or weak families. This would explain the permanence of similar differences in couples' fertility within each country. Moreover, in the strong-family areas, the anthropological structure centred on familist social norms is thought to have slowed the diffusion of new conjugal behaviour and - at the same time - encouraged very low fertility. As a result, in strong-family countries, fertility is, on average, lower than in weak-family countries. The apparent paradox mentioned at the beginning of this introduction could therefore be, at least partially, interpreted along these lines.

Gianpiero Dalla Zuanna, Department of Statistical Sciences, University of Padova.
Giuseppe A. Micheli, Institute of Population and Geographical Studies, Catholic University of Sacred Heart, Milan.

NOTES

1 In some regions of central and northern Italy, such as Liguria and Emilia-Romagna, the average number of children per woman between 1984 and 1996 was less than one. For data and comments on the evolution of fertility and other forms of marital and reproductive behavior in Italy in the second half of the 20th century, see the contribution of Rosina to this volume.

2 Actually, even within the individual countries, differences in fertility at the end of the 20th century are not completely in line with the 'classical' keys of interpretation. For example, in Norway and in the United States the probability of having a second and third child is higher among more educated couples.

3 Some recent studies have however shown that among Italian married couples, in which both partners are working, if the husband helps with child-care and housework in the period following the birth of the first or second child, the probability of having a second or third child is higher (Mencarini and Tanturri, 2003).

4 Data of Table 2 clearly show that among the elderly in southern Europe, contacts with family members are much more frequent compared with their peers in central and northern Europe. However, this does not necessarily mean that the elderly in the north feel more lonely or unhappy. In the same study cited here, the elderly were asked if they missed their children. The highest percentages of dissatisfaction were actually among those living in the countries of the south, where contact between elderly parents and children is more frequent. This finding, which is only apparently paradoxical, reinforces further still the idea of the profound differences in mentality, expectations and organization of family life between the countries of southern and central/northern Europe.

5 For example, Schneider and Schneider (1976), in one of the finest field surveys of social class history, wrote as follows: "the bipolar model of tradition vs. modernity has led to the holding of the peoples of the less industrialized regions of Europe, such as the south of Italy, as being responsible for their poverty. Thus, in the sixties,

Edward Banfield discovered that the inhabitants of the south of Italy were 'backward' because they were slowed down by a culture, which was too attached to the family. His book is a typical example of the dominant current of thought of the time (...). Our book presents an alternative reasoning to that offered by the school of Banfield, in so far as it seeks the reasons for economic underdevelopment in Sicily in certain relations of inequality (which) have made it possible for the island's wealth and resources to be exploited by outside groups over the centuries and which have simultaneously encouraged the growth of an unbridled and powerful system of patronage amongst Sicilians, with its power base ('clients') among rural entrepreneurs. Furthermore, we have refused to present Sicilian culture as 'traditional', demonstrating rather that there is a link between certain widespread values and these century-old relations of inequality..." Actually, the great merit of the Schneiders' study lay not in its stark contrast to Banfield's 'culturalist' approach, but in its researching of social and political co-variables (the emergence of those interstitial classes of *compradores* peculiar to peripheral capitalism, as discussed in Paul Baran's neo-Marxian approach, 1957) of the actual process of consolidation of an 'ethos'.

[6] For the typology of family models used here (Atlantic, Latin, Mediterranean), see the first contribution of Micheli. To be precise, in the Mediterranean model subordination is not to the family in the strict sense, but to the alliance of several family groups linked by blood ties (to the *nasab*, following Ibn Khaldoun).

[7] While such ideal-type categories are useful for the purposes of analysis, there naturally exists in reality a detailed range of intermediate conditions.

[8] Here I use a metaphor coined by Bourgeois-Pichat (1987). A 'holon' is "a particle which forms part of a whole", an independently organized entity in which two opposite tendencies take place: the one 'altruistic', placing it in contact with an element further up on the hierarchical ladder, the other 'selfish', corresponding to the individual's functions of self-sufficiency. Demography – continues Bourgeois-Pichat (1987) – "is the domain of holons *par excellence*. The couple, the family, the clan, the caste, the village, the city, the region and the nation are all holons located above the individual on the hierarchical scale. And an individual will have many different altruistic aspects".

[9] Ginsborg (1998) identifies an illuminating paradox in the changes in Italian society over the years 1980-1996: the reduced size of the Italian family was partly due to its own strength. "If individual strategies had been stronger and family ones weaker, it would have been possible to place the elderly in homes and persuade eighteen-year-olds to go out into the big wide world. But this attitude did not correspond to the Italian mentality".

[10] "Those who follow the familist rule are without morality only in the case of persons external to the family, but they apply the usual moral categories towards family members. From this point of view, those with no family are amoral individualists" (Banfield, 1958).

[11] From a theoretical point of view, the two models are now located at the two poles of the axis 'orientation towards Ego – orientation towards the collectivity' which Parsons (1951) initially included among his 'pattern variables', dimensions of alter-

native choice through which the social actor passes in order to give meaning to a situation. Another of the five pattern variables proposed by Parsons (1951) is provided by the axis 'universalism – particularism' (tendency to cope with a situation according to a general norm or according to *ad hoc* criteria, depending on the personal relation entertained between the individual and the situation). The familistic Mediterranean model is therefore properly a particularistic model, oriented towards the collectivity.

[12] Francesco Guicciardini (1483-1540) was a Florentine representative of the Medici first in Florence, then in the Papal States. Brought up amid the cultural atmosphere created by Savonarola and Marsilio Ficino, he was a friend of Machiavelli and achieved a top-level diplomatic career. His writings are the reflections of a disenchanted historian and politician. His career of work and reflection is comparable to that of Botero or Ibn Khaldoun.

[13] "The standing, which I have had with a number of popes, has made it necessary to love their grandeur for my particular (interest).." (Ricordi, third edition, n. 28). "Liberty is nothing else but the prevalence of laws and public orders over the appetites of particular men" (Discourse of Logrono, 1513). And further still: "do not take seriously those who preach liberty; I don't mean all of them, but there are very few exceptions (...) because almost all of them have more respect for their own interest" (Ricordi, second edition, n. 106). Note the strong similarity to the selfish dimension of amoral familism as described by Banfield (1958): "In a society of amoral familists, no one seeks to pursue the public interest, unless they have a private interest in doing so, and the law breaks down once the fear of punishment disappears".

ALESSANDRO ROSINA

FAMILY FORMATION AND FERTILITY IN ITALY

A cohort perspective

1. INTRODUCTION

In the last 30 years of the 20th century Italy has had a considerable decline in marital unions and in fertility levels (Barbagli, 1990; Delgado Perez, Livi Bacci, 1992; Golini, 1994; Micheli, 1995; Dalla Zuanna and Righi, 1999; De Sandre, 2000). It is now one of the countries with the lowest number of children per woman and the eldest age at first union. Another peculiarity – if we compare Italy to other western countries as a whole – is that despite these trends, the institution of marriage remains central to Italian culture. In fact, marriage is still the main reason for leaving the parental home and forming the first union, and out-of-wedlock births continue to be uncommon. Thus, families seem to be firmly founded on marriage, although fewer people are getting married, and fewer children are born out of those who are married. Is really this the picture of Italy facing the new millennium? The continuous rise in age at marriage and childbearing could suggest a deceptively pessimistic interpretation for both the period indicators[1] and for the total fertility of those generations still in the course of their marital and reproductive history. In assessing Italy's future, it is of crucial importance to understand whether new generations of Italians will compensate in later years for deficits in marriage and reproduction incurred earlier in life.[2]

It is also important to stress that the north of Italy must be considered distinct from the south. Studies not taking into account differences in their demographic behaviour do not properly show the possibility of recovery for the younger generations. In the north, the continuous decline in fertility has been chronically below the replacement level[3] for years (Santini, 1995). Rather than wondering if the low levels of fertility are temporary and whether a recovery of the replacement level will take place with the younger generations, it might be more productive trying to determine if the one-child family is actually establishing itself in the north as the main pattern. In the south, on the other hand, the levels of fertility have been much higher than

G. Dalla Zuanna and G.A. Micheli (eds.), Strong Family and Low Fertility: A Paradox?, 23-43.

in the rest of Italy for years. This is because the decline in fertility takes place after the birth of the second child. Will this decline continue? Will it affect the pattern in which a family generally has at least two children?

Finally, we would like to consider the role of informal unions. Are they going to remain at a marginal level in Italy or are they going to become more common? If their incidence increases, will this trend have a determining influence on family formation in the near future?

It is the intention of this study to raise some doubts with both those who have (as Umberto Eco would say) an 'apocalyptic' view, and those who have an 'integrated' view of the Italian demographic situation. The former emphasize the low levels of marital unions and childbearing and are sceptical about the future of the Italian population. The confidence of the latter on a recovery of the replacement level is based on a vision of convergence towards the standards of north-western Europe. As of now, it is difficult to see who might be right. We can only remark that the data available show an open history, with an unpredictable path, which from one generation to the next gives some answers but also raises some questions.[4]

The data we have here used refer to the female population[5] and are mainly from the survey "Family, Social Subjects and Childhood" (FSS98) performed by the Italian Statistical Office. This provides the most recent information on socio-demographic event history data. Where the source is not indicated, the information is based on our analyses of these data. Moreover, we have used some information taken from the International Fertility and Family Survey (FFS) project (UN-ECE, 1992; for Italy: De Sandre *et al.*, 1997), mainly for comparisons with other European countries.

2. WOMEN BORN UP TO THE EARLY 1960S

2.1. Leaving the parental home and entering the marital union

As is well known, the phenomenon of postponing marriage, and more generally of entering a union, began in the Scandinavian countries with the generations of the 1940s and has spread little by little throughout other western European countries. In northern Europe, the postponement of marriage has made some space for experiencing informal unions and autonomous choices. In Mediterranean Europe, on the other hand, this same phenomenon has almost always turned into a prolonged stay in the parents' house. In fact, in the majority of western European countries the age at leaving the parental home has remained stable at around 20 years old. In Italy and Spain this age is progressively being postponed towards the late 20s (Table 1).[6]

An high level of synchronization between leaving the parental home and marriage characterizes the transition to adulthood in southern Europe. However, with the women born in the 1960s in the north-centre of Italy something starts to change (Table 2). The percentage of women who leave for other reasons and only subsequently enter a conjugal union, which historically is at around 8% in northern Italy, for the first time rises beyond 12%. This is mainly due, as we shall see later, to the beginning of the spread of informal cohabitations. This is not the case of southern Italy, however, where the polarization between leaving in order to get married and an indefinite stay at the parental home is still very strong. The rate of women who don't marry remains nevertheless well below 15%.

Table 1. Median age at leaving the parental family and at first union. Women.

	Born in the early 1950s		Born in the early 1960s	
	Leaving parental home	First union	Leaving parental home	First union
Austria	20.0	21	19.1	20.7
Flanders (Be)	21.2	21.4	21.7	22.3
France	20.3	21.4	20	21.7
West Germany	n.a.	n.a.	21	23
Netherlands	19.6	20.0	19.5	21
Norway	n.a.	n.a.	20.2	21.1
Spain	23.2	23.2	>23.5	24
Italy	22.2	22.5	23.8	24.2

Source: FFS (Corijn, 2000).

Table 2. Leaving the parental home and marital status (%), per generation. Situation at 35 years old. Women of north-centre and south-islands.

	North-centre				South-islands			
	Married		Unmarried		Married		Unmarried	
	Directly	After leaving	Living with parents	Out of the parental home	Directly	After leaving	Living with parents	Out of the parental home
1930-39	82.1	7.4	6.8	3.8	83.6	3.3	10.0	3.2
1940-44	83.6	8.2	5.2	3.1	81.5	5.3	10.3	2.9
1945-49	83.7	8.6	3.3	4.3	83.8	4.3	8.8	3.1
1950-54	83.2	7.5	4.2	5.1	84.4	4.0	8.5	3.1
1955-59	79.0	8.1	7.9	5.1	84.3	5.2	8.1	2.5
1960-62	77.8	12.1	6.6	3.6	80.1	5.2	11.1	3.7

Figure 1. Distribution by age at first marriage per generation. Women in the north-centre.

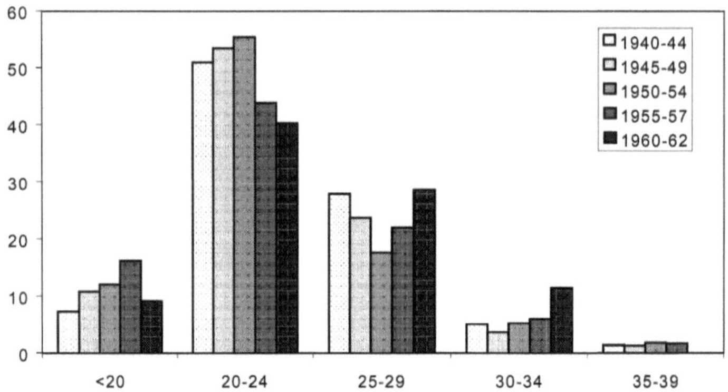

In Italy, the progressive lowering of age at marriage reaches its highest peak with the cohorts born before the 1950s. With the birth cohorts of the late 1950s we can notice the emergence of two opposite behaviours: the increase of precocious marriages (under 20 years old), which is in this period at its highest peak; and the beginning of the trend of postponing marriage after 25 years old. The most striking finding in comparison to the previous generations is the considerable decrease in marriages in the age group of 20-24 (Figure 1).

Figure 2. Distribution by age at first marriage per generation. Women in the south-islands.

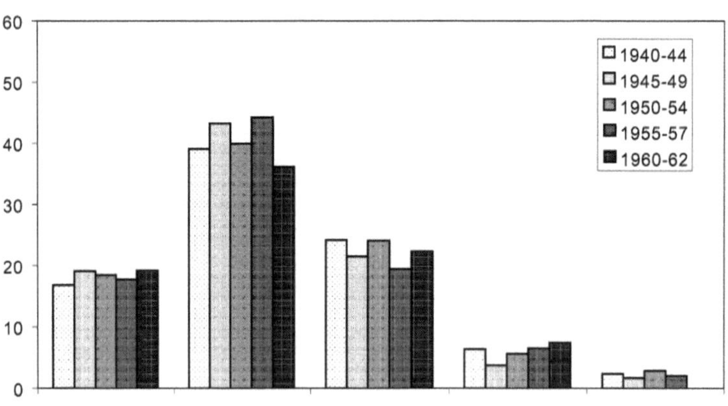

The turning point is therefore at 20 years old for the birth cohorts of the late 1950s. This generational behaviour affects the evolution of the (period-based) total marriage rate. This rate changes in fact from 942‰ to below 700‰ in 1985. The decrease of marriages in the 20-24 age group of the birth cohorts of the late 1950s can clearly been seen between the second half of the 1970s and the first half of the 1980s. It is only at the end of the 1980s that an increase in marriages after 25 years old has a beneficial compensatory effect on the period indicators. This effect is obscured however by the continuous process of marriage postponement in the younger generations. None of this applies to the south, where the differences in demographic behaviour among the generations observed are very limited.

2.2. Motherhood and marital status

The importance of marriage in Catholic countries as a prerequisite for building a family is clearly testified by the values in Table 3. If marriage continues to be a necessary condition in Italy to have children, it proves however to be less and less a sufficient one. With the women born in the 1950s we observe in fact an increase of married women without children in the north-centre (Table 4). These are often women with a high level of education. Generally we can say that in the north-centre, starting from the women of the birth cohorts of the 1950s with a higher education, there is a clear weakening in the relationship between marriage and reproduction (Table 5). This also includes a considerable amount of women having children out of wedlock. In the south, on the other hand, there is an increase in the number of married women with children. This can be seen as a result of a greater tendency to both getting married, and having children when married. This could be linked to a fall in male emigration, which reached particularly high levels with the birth cohorts of the early 1940s.

Table 3. Partnership status at first birth. Women interviewed at FFS (age 25-29).

	Married	Consensual union	Single
Austria	50.3	24.6	25.1
Belgium	91.1	5.2	3.7
France	59.9	28.1	12.0
West Germany	63.4	9.1	27.5
Netherlands	83.0	11.0	6.0
Norway	61.3	25.5	13.2
Spain	87.8	5.3	6.8
Italy	90.8	4.3	4.9

Table 4. Distribution (%) by marital status and children. Situation at 40 years old. north-centre and south-islands.

	North-centre				South-islands			
	Never married		Married		Never married		Married	
	Without children	With children	Without children	With children	Without children	With children	Without children	With children
1930-39	7.3	1.0	5.8	86.0	10.9	1.1	6.2	81.9
1940-44	5.7	1.3	4.2	88.8	9.1	1.9	6.2	82.8
1945-49	5.4	1.5	4.8	88.3	8.6	2.1	2.8	86.6
1950-54	6.1	1.5	6.7	85.7	7.9	1.0	3.9	87.2
1955-57	8.4	1.5	7.7	82.4	8.4	1.4	2.9	87.3

Table 5. Distribution (row %) by marital status and children, per level of education. Situation at 40 years old for the birth cohorts of 1950-57. north-centre and south-islands.

	North-centre				South-islands			
	Never married		Married		Never married		Married	
	Without children	With children	Without children	With children	Without children	With children	Without children	With children
Degree	11.4	4.3	15.1	69.2	10.9	1.4	4.0	83.7
High school	9.5	0.9	6.6	83.1	7.6	1.1	3.7	87.6
Less	4.6	1.4	5.9	88.1	7.7	1.2	3.3	87.8

Table 6. Median age at first child. Women.

	Born in the early 1950s	Born in the early 1960s
Austria	22.8	24.0
Great Britain	25.3	27.7
Flanders (Be)	24.2	26.4
France	24.0	25.3
Netherlands	25.0	28.0
Spain	25.3	26.8
Italy	24.8	27.2
North	25.8	28.6
South-islands	23.8	25.4

Source: FFS (Corijn, 2000; De Sandre *et al.*, 1997).

2.3. *Age at first child*

Italy as a whole shows two very distinct realities which clearly emerge from the timing of entry into motherhood. As can be noticed in Table 6, northern Italy reaches the highest levels in Europe in the postponement of the first child. Among the younger generations almost 45% of women start bearing children after 30. The south, on the other hand, is at the same level of those western countries with the earliest age at entry into motherhood. The process of postponing fertility proves to be nevertheless widespread.

The increase in the levels of fertility after 30 years old partly makes up for the decline in the previous age groups. In the south the birth cohorts of the late 1950s highlight a large increase in early fertility, which however decreases considerably with the following generations (Figures 3, 4).[7]

Incidence of a recovery of fertility after 30 years old seems to become noticeable with the birth cohorts of the 1960s only. In comparison to the previous generation, the women in this age group almost completely make up for the low levels of first order fertility shown before 25 years old. In the following generations, early fertility further decreases.

Less than 20% of the birth cohorts of the late 1960s have a child before they are 25 years old. It therefore remains to be seen whether the most recent generations will show an ability to make up for such low levels of fertility with an equally higher increase at an older age. The probability of a pregnancy decreases slowly from 25 to 35 (McDonald *et al.*, 2003).

The choice of postponing the birth of the first child from the late 20s to the early 30s does not seem to have any relevant consequence on the risk of having a pregnancy from a biological point of view. However, if this means to defer the birth of a second child after 35 years old, many women could be compelled to renounce the desire of enlarging the family beyond the first child (Rosina and Colombo, 2003).

2.4. *Total fertility*

The period total fertility rate is affected by the postponement of childbearing. As shown earlier, in Italy this process has begun with the birth cohorts of the late 1950s. Similarly to the marital unions (see section 2.1.), the considerable fall in births before 25 years old is visible between the late 1970s and the early 1980s. The total fertility rate goes from over 2.2 in the early 1970s to just above 1.2 in the late 1980s - a statistic internationally unrivalled. The beneficial effect of a recovery in the fertility levels after 25 years of age begins to be noticeable on the period fertility only towards the late 1980s. In the 1990s the total fertility rate remains around 1.2 (Figure 5).

Figure 3. Distribution (%) by age at first child, per generation. Women in the north-centre.

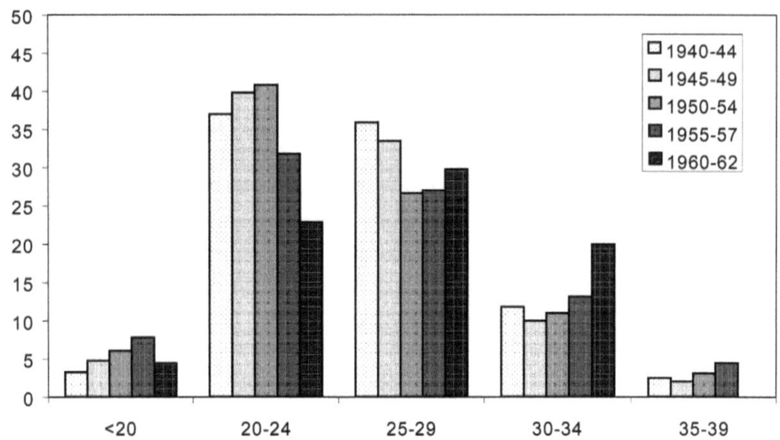

Figure 4. Distribution (%) by age at first child, per generation. Women living in the south-islands.

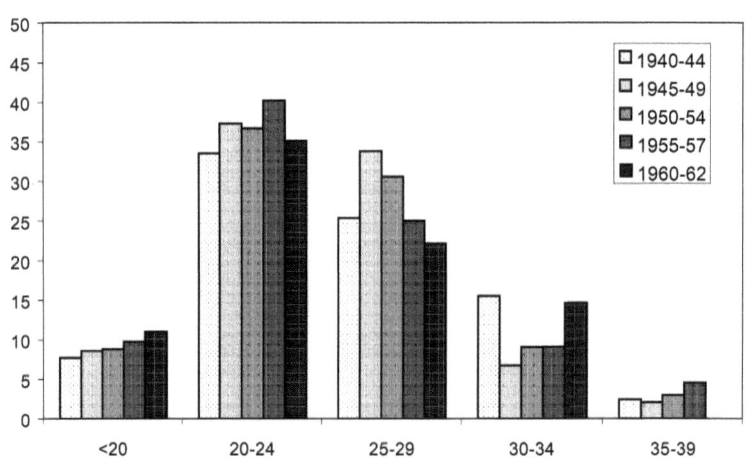

Figure 5. Lexis diagram. Representation of the timing of the process of postponement (A) and of recovery (B)

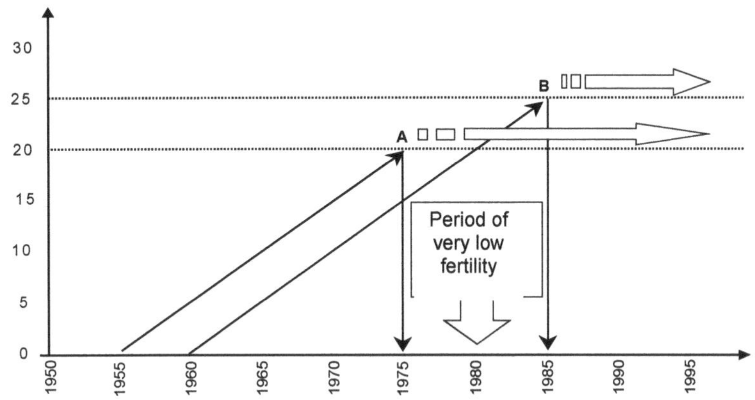

In order to understand to what extent the decrease in fertility before 25 years old corresponds with a change in reproductive behaviour, which could therefore affect permanently the future of the Italian population, we must refer to the generational behaviour. Frejka and Calot (2001) sum up in a table the overall level and the generational trend in the total cohort fertility rates for more than 20 countries.[8] As with the other countries listed, the data relating to Italy refer to the whole of the country. To these data we add the abbreviations NC for north-centre and SI for south-islands (Table 7). Once again, in comparison to the levels of fertility of other western countries, the north-centre and the south-islands show extreme behaviour on opposite poles. In the north, the total fertility rate has been below the replacement level since the 20s.[9] With the birth cohorts of the early 1960s it reaches just below 1.5 children per women. The TFR in the south, on the contrary, has always been close to the levels of the most prolific western countries, although it shows a remarkable decrease from about 3 children for the cohort of 1931 to about two for the cohort of 1962.

In the south-islands (Figure 7) we can observe a gradual establishing of the pattern of the family with two children (Santini, 1995). The decline in total fertility (Table 7) is due, at least until the 1950 generation, to the decrease of families with 3 or more children. In the north-centre, the cohorts of the post-war period mainly show, on the other hand, an increase of single women. This, together with the decrease in families with 3 children, brings the total fertility rate to about 1.5 children per woman (Table 8).

It doesn't seem that the pattern of the one-child family is establishing itself, at least for the cohorts before the 1960s (Figure 6).

Table 7. Number of countries in which women born in 1931 1946 1962 have experienced specified total cohort fertility rates (TCFR). Italy is subdivided into north-centre (NC) and south-islands (SI).

	Birth cohort		
	1931	1946	1962
Less than 1.60	□	□	2 (NC)
1.60-1.79	□	2	6
1.80-1.99	(NC)	10 (NC)	9
2.00-2.19	7	7	8 (SI)
2.20-2.39	6	5	2
2.40-2.59	3	2 (SI)	□
2.60 and above	7 (SI)	□	□
Total number of countries	23	26	27
Average TCFR (un-weighted)	2.48	2.08	1.89

Source: Frejka and Calot (2001). Our addition for the north-centre and south-islands of Italy.

Table 8. Total cohort fertility.

	North-centre	South-islands
1930-39	2.02	2.73
1940-44	1.94	2.49
1945-49	1.70	2.43
1950-54	1.63	2.19
1955-57	1.51	2.21

It is not easy to speculate what might happen with the next generations. If in the south the pattern of the family with at least two children persists, we can foresee a further decline in fertility levels - especially after the second child - although a convergence towards the levels reached today by the north-centre is very unlikely. On the other hand, if in the north-centre the pattern of the one-child becomes widespread (though there is no evidence of this until the cohorts of the late 1950s) then the fertility levels will be falling further.

Figure 6. Distribution (%) by number of children, per generation. Women in the north-centre.

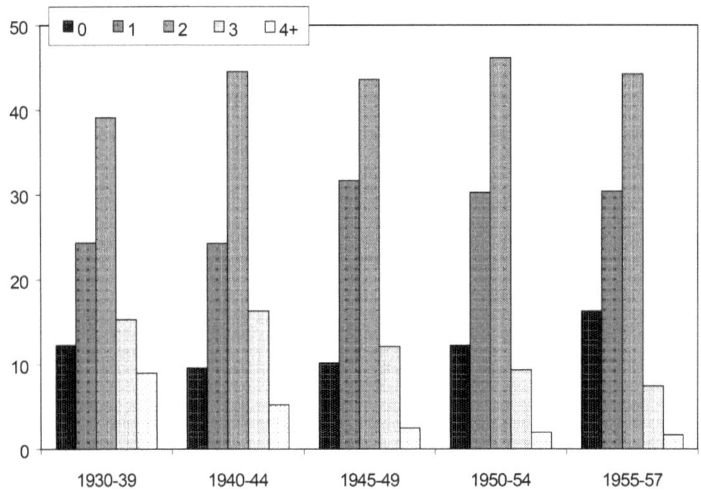

Figure 7. Distribution (%) by number of children, per generation. Women in the south-islands.

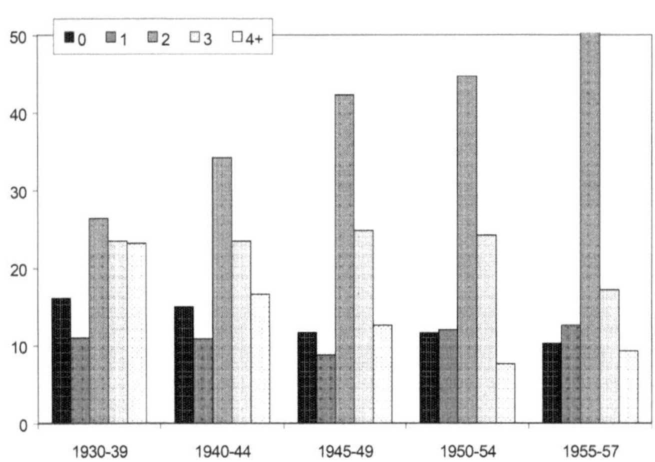

In calculating the number of children that the birth cohorts of 1960-62 had before reaching 35 years of age, we notice that in the north-centre, the women who had one child are 35%: about 5% more than the previous generation; while the women with two children are 39%: about 5% less than the previous generation. This is the generation which highlighted a remarkable increase in first order fertility in the age group 30-34 in comparison to previous generations (Figure 3). It is therefore plausible to expect a considerable amount of second children to be conceived after 35 years old. But the uncompleted information regarding the generations still in the course of their reproductive history at the moment of the survey shows a further postponement of both marriage and childbearing after 25 years of age. It is therefore reasonable to foresee that for these generations there will be an increase of women with less than two children. It is too early, however, to state that there is a firm tendency towards an increase in the number of families with only one child.

Figure 8. Distribution (%) by number of children, per level of education. Women born in the 1950s in the north-centre

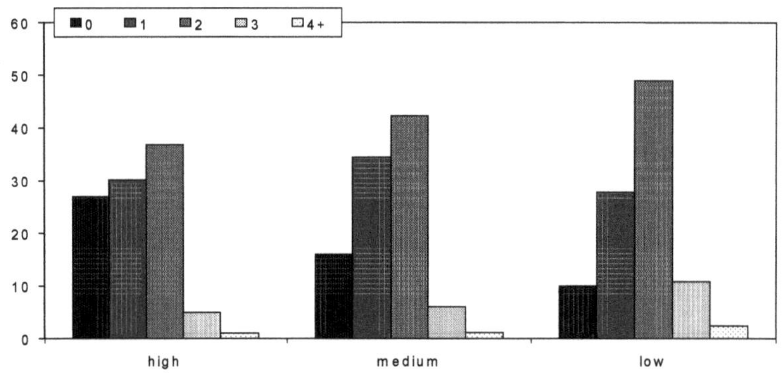

As of now the only definite information is that there is a relevant postponement of marriage and childbearing, and a clear tendency to compensate to this after 30 years old. It is too soon to see whether the most recent generations will be able to make up for the further decrease in fertility before 25 years old.

2.5. Number of children and level of education

Taking a deeper look at the relationship between number of children and level of education we perceive a likely further decrease in the fertility levels. About 55% of the women born in the 1950s with a high level of education have at most one child (Figure 8). A more thorough study of the transition to the second child shows that the level of education has a positive effect if the comparison takes into account the age at first child. In the north, women with a high level of education get married at an older age – both because of university studies and career commitments. After taking this initial disadvantage into consideration, these women show a greater ability in managing time for their career and for the care of their children.[10] Indeed, if we take into account occupational status we obtain an even higher positive effect of the level of education (Rosina, 1999).

These results seem to suggest that for these women the tendency towards having only one child could be more a case of having to forgo the desire of having a second child than a real choice.[11] This is probably due to a combination of commitments imposed by a professional career and the older age at the time of marriage. Further support to this interpretation comes from the fact that this tendency occurs mainly among women with a university degree who come from a family with poor financial and cultural resources (Figure 10).

Figure 9. Distribution (%) by number of children, per level of education. Women born in the 1950s in the south-islands.

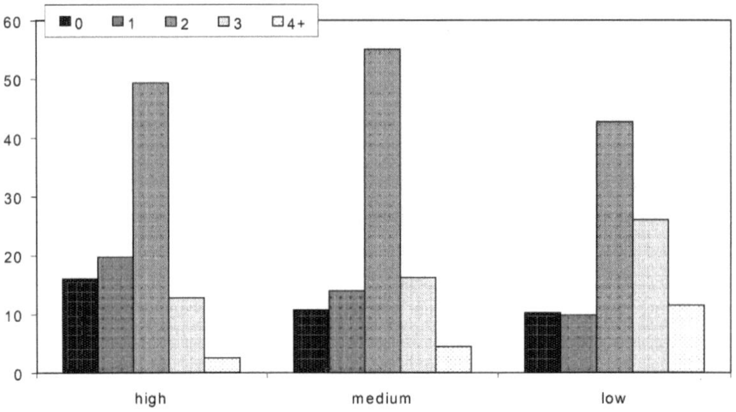

Figure 10. Distribution (%) by number of children, per level of education of the father. Women with a degree. Cohort 1950-57, north-centre.

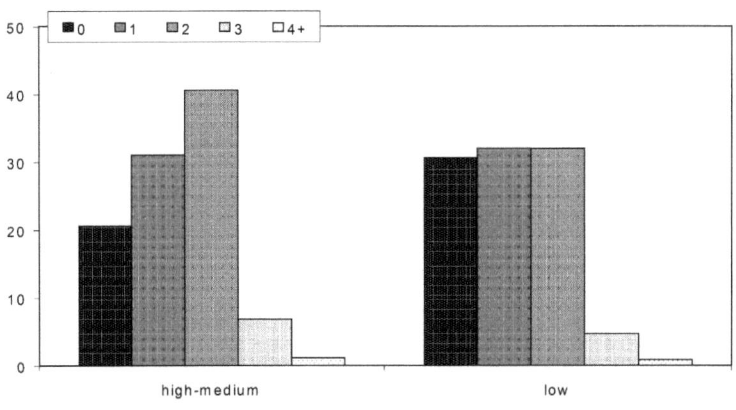

The same result could also be read from another perspective. Besides the combination of having high professional expectations and poor family resources at the origin, the fall in fertility could be attributed to a different personal attitude. In comparison to the women coming from a more educated parental family, these women, according to Micheli (1999a), are characterized especially by a combination of material priorities and family values, which would promote a less favourable attitude towards procreation. Maybe these women have a more familistic orientation. Perhaps, they are closer than others to strengthening vertical relations and social mobility with the choice of investing in only one child (Dalla Zuanna, 1995; Dalla Zuanna and

Righi, 1999). This interpretation seems to be supported by the results from a study on the relationship between parents and children (Allegra and Rosina, 2001), where it is observed that parents with a high level of education coming from a low socio-economic status are more likely to invest in only one child.

3. THE BEHAVIOUR OF YOUNGER COHORTS

We have so far analysed the data concerning the behaviour of the birth cohorts who have already concluded their reproductive history. Let us proceed to consider the younger generations. We can speculate, in relation to the partial biographical information observed with the 1998 survey, if and in what potential ways their behaviour will be different from that of the previous generations.

We have seen that the postponement to an older age of both marriage and childbearing begins with the birth cohorts of the late 1950s and becomes more significant with the birth cohorts of the 1960s.[12] In comparison to the previous generations, women born in the 1950s in the north-centre are characterized by a marked renunciation of marriage and of having children. These are mainly women with a higher education, particularly those coming from a parental family with poor financial and cultural resources.

Higher personal expectations, developed alongside higher education, combined with a traditional environment – as to family patterns, gender roles, and working time management – could have almost completely polarized work and family for these women. According to Piazza (1995), this generation is typified by a 'daily synchronicity', or in other word by the effort of equally managing and maintaining the two poles of family and work. This would create a situation filled with hard work, difficulties, lack of social resources, and hostility from the cultural environment. The obvious way out of this situation is the sacrifice of one of the two poles. It is only with the birth cohorts of the 1960, in their reproductive period in the late 1980s, that a 'diachronic strategy' seems to emerge, characterized by a postponement and recuperation of childbearing and by a greater flexibility in the management of working time.

With the birth cohorts of the early 1960s there is an increase in marriages if compared to the cohorts of the late 1950s. The decrease in direct marriages is compensated by the marital unions entered after a period of independence from the parents (Table 2). Moreover, a recovery of fertility after 30 years old seems to gain strength. The birth cohorts of the 1960s therefore show some signs of decrease in the rigidity of female biographies, combined with an increase in the possibility/ability of managing time more efficiently.

According to Micheli (1999a), the process of postponement of these events may be attributed to a sense of insecurity which pushes young adults to defer making decisions perceived as particularly binding or requiring a considerable commitment. Micheli makes a distinction between anxious insecurity and avoidant insecurity. The latter leads to a "declared collapse of the procreative project". With the anxious insecurity, on the other hand, the postponement is due to the higher value placed on marriage and children. The women in this group tend to postpone these choices rather than reject them. They generally live in metropolitan areas and tend to have post-materialistic values. Moreover, this specific group seems to have the attitude and the financial and cultural resources needed for the fulfilment of the procreative project and the maintenance of a diachronic strategy sustaining both high professional expectations and a family with more than one child.

Figure 11. Percentage of women having experienced a first union before 25 years old and cohabitations as percentage of first unions. Birth cohorts of 1966-70 in some of the countries in western Europe.

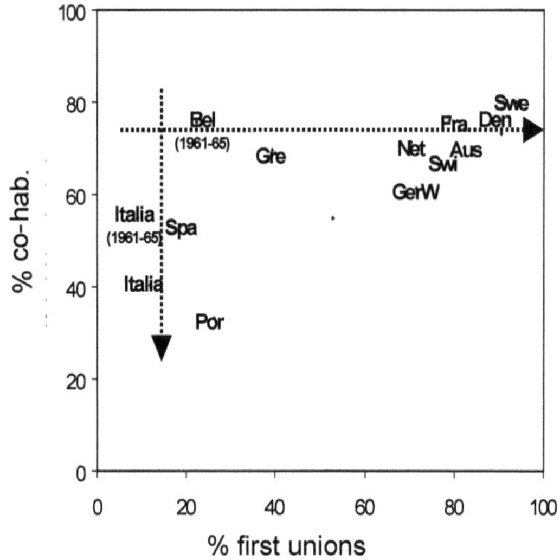

Source: FFS data

We can also speculate that this sense of insecurity will make informal unions more popular among the younger generations, pushing them towards less binding choices, more options, and a better flexibility in the management of time. The popularity of informal unions could encourage an early exit from the parental home, and eventually the formation of a family. It is interesting to observe (Figure 11) that European countries with a higher level of informal unions at a young age tend also to be those in which the role and value of marriage has been questioned more. From this point of view, the Catholic countries in southern Europe show an unusual behaviour. The lower popularity of informal unions (horizontal evolution in the Figure) is linked to a fall in the forming of unions at an early age (vertical evolution in the figure). It is understood that an early union favours fertility[13]. Indeed, the countries in which the informal unions are more popular tend to also have higher levels of fertility (Dalla Zuanna, in this volume).

Although the number of informal cohabitations is relatively low in Italy there is a clear and rapid increase of this phenomenon in the north-centre, not only in the cities but also in the countryside (Figure 12). This does not mean, however, that there will necessarily be a convergence towards the levels of other western countries. This would be a hasty prediction, primarily because the process of leaving the parental home have a completely different meaning in Italy - and generally throughout the Mediterranean area - than in the countries of northern Europe. Such a choice is closely dependant on the strong relationship between parents and children, a tie that is anthropologically rooted in Italian society (see Reher and the first chapter of Micheli, in this volume). This relationship is essentially fed by the marked material and emotional involvement of the parents in the lives of their children (not only during childhood, adolescence and youth, but also during adulthood). This is based on the assumption that the children's success (and especially their well-being) is recognized (at least in part) as a consequence of the parents' sacrifices and of a far-sighted family strategy. This attitude is defined as 'familist' by some authors (Dalla Zuanna, in this volume).

Particularly, the resources of the parental family have a crucial role in the development of young Italians, not only in terms of tools offered as initial endowment (specifically in the case of formative opportunities) but also subsequently as support and assistance at each stage (house, employment, children, and so on), and of prompt intervention at each difficult moment.[14] All this is reinforced by a substantial lack of 'state support' in the crucial events of the life of the Italians, which at the same time is cause and effect of the crucial role of the family in Italian society (De Sandre, 1997).

The fact that parents see their children as extensions of themselves implies, besides a keen investment on them, also a heavy conditioning of their young children's choices. The specificity of the Italian family, which is to say the continuity of strong vertical ties, can be compatible with a non-traditional choice made by the children only in the condition that such a choice does not clash irremediably with the parents' values and is socially acceptable in the context of origin. Only in this case can the innovative behaviour of children be carried out without jeopardizing the (physical, affective, instrumental) closeness with their parents.[15] This especially happens if the informal union takes the shape of a pre-marital cohabitation (as it happens in the great majority of the cases in Italy).

We can expect, therefore, that the opening up to such a behaviour, although with a delay of some generations in respect to other western countries, depends not only on the children's cultural capability of making non-traditional choices, but also (and above all) on the cultural availability of the parents in accepting that their children can make those choices.

Several empirical analyses seem to support this hypothesis (Rosina, 2001; Rosina *et al.*, 2003; Rosina and Billari, 2003; Barbagli *et al.*, 2003). Cohabitations have started to spread, mainly with those born in the second half of the 1960s. Before this generation, informal unions emerge almost exclusively in less traditional contexts (North-centre and metropolitan areas), and are experienced by those young people who are less conditioned by the socio-cultural context in which they live (with a medium-high educational level, and low attendance to religious services). But even more important is the fact that the characteristics of the parents rather than those of the very children seem to have a determinant influence on such a choice (Figure 13).

Figure 12. Percentage of cohabitations as a form of first union, per generation and municipality size. north-centre. Cohorts 1960-67

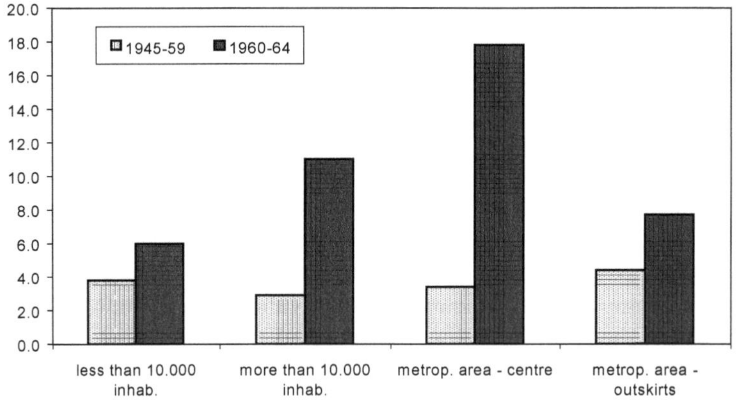

Figure 13. Percentage of cohabitations as a form of first union, according to the woman's and the father's level of education. Women of the north-centre. Cohorts 1960-67

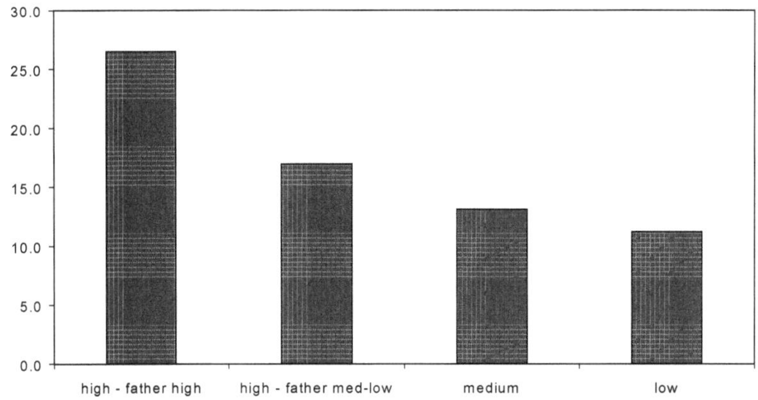

4. CONCLUSION

The results here described don't confirm the conceptualization of Italy with an irreparable negative anomaly, nor the image of a latecomer in the path already marked out by other western countries.

It is clear that Italy differs from other European countries for a continuous significant decline in marriages and procreation before 25 years of age. Yet it is also true that the recuperation in 30 year-olds in the north-centre seems to make up almost completely for that fall.[16] Although the number of informal unions is still relatively low, they are rapidly increasing. And this could eventually encourage the formation of early unions, which would favor an increase in fertility. Considering the anthropological differences (Reher and the first chapter of Micheli in this volume) in the process of autonomy of the children, it is hard to believe that a widespread increase of informal unions could in the near future push the current rate of Italians leaving the parental home before age 25 from about 1 out of 3 towards the northern Europe rate of approximately 3 out of 4. It is more likely that the young generations will continue leaving the parental home relatively late, albeit perhaps a bit earlier than the previous generations. And this would still be compatible with a fertility rising toward the average rate of two children per woman.

Establishing of the pattern of the family with one child seems more controversial than generally considered, mainly on the basis of period data. It is very likely that in many cases Italians intend to have more than one child, yet must forgo the second one. This is what happened with the birth cohorts of the 1950s with a high socio-cultural status but coming from a parental family with low financial and cultural resources. And this is what could still happen to some of the women of the younger generations who postpone childbearing beyond 35 years old.

If with the younger generations in the north the level of fertility does not seem to be decreasing to one child per woman, the progressive postponement of the onset of family formation allows with difficulty for such a recovery to reach the replacement level. A clear and radical change of direction in the level of fertility can only be achieved through a revival of unions at a young age. The increasing popularity of premarital cohabitations could be a sign of movement in this direction.[17]

Alessandro Rosina, Institute of Population and Geographical Studies, Catholic University of Sacred Heart, Milan.

NOTES

[1] "By extrapolating current low levels of fertility into the future, analysts often unwittingly ignore the fact that the rates are temporarily depressed by a rising age at childbearing. Eventually, the age at childbearing will stop rising and the removal of this fertility-depressed effect might well result in a rise in the TFR, as in fact happened in the United States in the late 1980s" (Bongaarts and Feeney, 1998).
[2] For example, the marked decrease of marriages before 25 years old has lead several analysts to predict that over 20% of women born in the early 1960s would always remain single. This prediction, considering the most recent data, excessively emphasizes the consequences of the decrease of marriages before 25 years old without taking into account the possibility of recovery after 30 years old.
[3] As is well known, this corresponds to about 2.1 children per woman.
[4] Old generations transmit tools and rules elaborated according to their own time, while the new generations re-elaborate them creatively according to the constraints and opportunities pertaining to the present time.
[5] In Fabroni *et al.* (2001), we can find a male perspective analysis of the formation of the family, with a special focus on the transition to the first child.
[6] For Italian women born at the end of the 1960s, the median age at leaving the family of origin proves to be later than 26 years old. While in France and in Scandinavian countries, for example, it is stable at around 20 years old.
[7] Early childbearing can be reconstructed for all the women who were over 20 years old at the time of the FSS survey. We can therefore consider the birth cohorts until the early 1970s. In the south, the chance of having the first child before 20 years old

is clearly on the decrease starting with the cohorts of the middle 1960s, reaching below 5% with the birth cohorts of the 1970s. In the north-centre, on the other hand, the chance is already below 5% with the generations born at the beginning of the 1960s.

[8] Mainly European countries, with the addition of the US, Japan, Australia, and New Zealand.

[9] In the north-west the total cohort fertility rate is equal to less than two children per woman already with the birth cohorts of the late 19th and the early 20th century (Livi Bacci and Ventisette, 1972).

[10] Kohler *et al.* (2001) also formally show that a high ability of accumulating human capital can at the same time lead to a postponement of the first child and to an increase of the total fertility rate, controlling for the age at first child. For more general observations on the relation between age at first child and total number of children, and on the interaction of this relation with the social context see Billari and Rosina (2003).

[11] Very few women would declare wanting just one child when still at the beginning of their reproductive history. This is also the case of well-educated women living in the north-centre (De Sandre *et al*, 1997, p. 161). We must consider though that when giving such an answer many women tend to simply give a number of children they consider to be socially acceptable. Even so, we would have an idea of what kind of family size would be socially desirable and eventually questioned.

[12] So much so as to have been defined a syndrome (Livi Bacci, 1997; Livi Bacci and Salvini, 2000).

[13] Billari and Rosina (2003). See also Baizan *et al.* (2001) for Spain.

[14] The help of the parents proves to be fundamental both as a protection from the risk of possible difficulties, and as a resource to come out of those difficulties (Rosina, 2001).

[15] One of the main Italian specificities is the high residential closeness between parents and married children (Barbagli, 1997; Tomassini *et al.*, 2003), which is instrumental in the affective and material support of the children even when they have already left the parental home. Such a closeness neutralizes the possibility of a 'tolerant indifference' or of a 'reflexive ignorance' among the generations (Beck, 1997) which would enable young people to make free choices even though parents do not approve of them.

[16] It is very likely that the cohorts of the 1960s will have an average number of children inferior to the cohorts of the late 1950s, even though we can expect a deceleration in the decrease with the younger generations, mainly because of their greater ability for recuperation.

[17] In recent years in the metropolitan areas of the north-centre we have observed both an increasing popularity of premarital cohabitations and a recuperation of fertility (Dalla Zuanna and Crisafulli, 2001). See also Billari and Rosina (2003) for an individual level analysis of the Italian data on the relationship among age, typology of first union and total fertility.

DAVID S. REHER

FAMILY TIES IN WESTERN EUROPE

Persistent contrasts

1. PREMISE

In the Western world it is not difficult to identify areas where families and family ties are relatively 'strong' and others where they are relatively 'weak'. There are regions where traditionally the family group has had priority over the individual, and others where the opposite has tended to happen, with the individual and individual values having priority over everything else. This strength and resilience of family loyalties, allegiances and authority can be seen most clearly within the co-residential domestic group and among persons from the same conjugal family, though they extend to the larger kin group as well. These differences may well have characterized the European family for centuries and there are few indications that they are changing today. The way in which the relationship between the family group and its members manifests itself has implications for the way society itself functions. Politicians and public planners would do well consider the nature of existing family systems when designing certain social policies.

The geography of these 'strong' and 'weak' family systems does not appear to follow the classic division of Europe into stem and nuclear-family regions. The dividing line, in some ways, is actually much simpler, with the center and northern part of Europe (Scandinavia, the British Isles, the Low Countries, much of Germany and Austria),[1] together with North American society, being characterized by relatively weak family links, and the Mediterranean region by strong family ties.[2] The specific boundaries of different family systems are often not crystal clear and there is much subregional difference. For example, in some respects Ireland does not fit well into northern European family patterns, there are indications that northern and southern France often walk divergent paths, and the southern fringes of Spain, Italy or Portugal often show distinct characteristics from the northern parts of those same countries. Within individual societies, there is also much room

G. Dalla Zuanna and G.A. Micheli (eds.), Strong Family and Low Fertility: A Paradox?, 45-76.

for heterogeneity affecting families and family life. This multiplicity of forms and behavior, however, does not negate the existence of more general regularities affecting large areas of Europe.

For the most part, our analysis does not include the Europe lying to the east of John Hajnal's famous Leningrad-Trieste line which set apart fundamentally different marriage regimes, demographic structures and family systems on the European continent (Hajnal, 1965, 1983). There, forms of familial organization are sufficiently different to warrant their own specific study. While limiting the contexts of our analysis, this enables us to keep comparisons fairly straightforward. Contextualizing present behavior patterns in the light of historical experience, however, can and should be used when examining family systems in Eastern Europe or in any other society.

In this essay, then, the general has been stressed over the specific; the big picture set out in bold strokes has prevailed over attention to detail. In so doing, we hope to portray basic contexts of comparison and underscore the key issues involved as clearly as possible. This entails a certain inevitable degree of reductionism affecting the heterogeneity of the Europe experience within and across societies, as well as over time. In the future, additional reflection and work will go far towards filling in many of the gaps in our argument.

2. FROM THE PRESENT TO THE PAST

Vestiges of these divisions can be seen quite clearly in many aspects of family life. Among the most important are those centered on the moment of transition when young members of the family group set up households of their own or in the way in which the family organizes support for its most vulnerable members.[3] In northern Europe and in the United States, young adults normally abandon their parental households when they have acquired a degree of maturity so as to start out their adult lives on their own, lives that are occupied by their studies or by establishing of economic independence from their parents. Their jobs, which are often unstable or even seasonal, might also enable them to save for their own marriages, though nowadays this sense of saving is much less important than their effort to settle into an independent life. Often these initial forays into the adult world are done while sharing housing with friends and colleagues who are experiencing similar moments of their own lives. Later, often years later, these young people marry and once again start a new household, though this time with the intention of founding a family within the context of a stable relationship with another person.

In societies of Mediterranean Europe, the process of leaving the parental household is quite different. In these societies, the definitive departure of young people generally tends to coincide more or less closely with their marriage and their finding a stable job. The years between adolescent maturity (ages 18-20 years) and the age at marriage are spent fundamentally within the parental household. If a person gets a job during this period, he normally will continue to live at home, a strategy that enables him to save for his or her own marriage. Generally marriage does not even enter the picture unless it is accompanied by the corresponding emancipation from the parental home and the formation of a new household. This entire process is aptly crystallized in the traditional Spanish aphorism *casada casa quiere*.[4] In this way, in Spain and in many other southern European countries, a stable job, access to adequate housing, leaving the parental household, and marriage tend to be closely intertwined events. In fact, an excellent indicator of the labor market and unquestionably the best one for the rate of family formation in southern Europe would be the incidence of first marriages among young adults.[5]

In both contexts there are, of course, many exceptions. In England, Holland, and the United States, for example, there are numerous instances of young adults remaining at home past 20 years of age, and in Spain and Portugal some people leave home before marriage and others continue to live with their parents after marriage, at least for awhile. In fact, temporary coresidence of parents and married children, and even prolonged periods of economic help, have never been infrequent, either in the past or today.[6] Nevertheless, these moments of help were always considered as exceptional by everyone. These exceptions only underlie the great differences between northern and southern Europe on this point.

These divergent practices appear to have deep historical roots. From at least the latter part of the Middle Ages until the second half of the nineteenth century and the early years of this century, it was very common in rural England for young adults to leave their parental households to work as agricultural servants in other households for a prolonged period.[7] Servants might go to households of higher social and economic standing, though servant exchange among households of the same social status was widespread. In other words, it was common for a farmer to send his own son out as an agricultural servant in a farm, say, in the neighboring village, while he took other young servants into his own household as agricultural laborers. This practice appears to have affected the majority of the young adults in rural England during the seventeenth and eighteenth centuries. Peter Laslett has pointed out that approximately half of all young people of both sexes between 15 and 24 years of age were servants.[8] According to Kussmaul (1981), in a large sample of English communities between the seventeenth and the eighteenth centuries, about 60 percent of all farmers had servants, and these made up about half of the supply of non-family labor in rural areas and accounted for 10-12 percent of the total population. Implied by the extent of this practice is the fact that the great majority of young adults in England left their parental households more or less permanently between 15 and 19 years of age.[9]

Table 1. Percentage of servants in several northern and central European populations.

Country	Sample or place	Date	Percent servants
Denmark	Sample of parishes	1787/1801	17.6
Iceland	3 counties	1729	17.1
Norway	3 areas	1801	8.9
Belgium	9 Flemish villages	1814	14.2
Austria	Large sample (19 listings; median)	17-18th centuries	13.0
Netherlands	4 places	1622-1795	11.7
Germany	Grossenmeer	1795	10.7
France	Longuenesse (north)	1778	12.6
France	2 southern villages	1644-1697	6.4

Sources: Denmark (Hajnal, 1982; Johansen, 1975); Iceland (Statistical Bureau, 1975); Norway (Drake, 1969; Hajnal, 1982); Belgium (Wall, 1983b); Austria (Schmidtbauer, 1983); Holland (van der Woude, 1972; Laslett, 1977a; also Schellekens, 1991); Germany (Laslett, 1977); France (Laslett, 1977a).

There is ample evidence of the importance of servants in other northern European societies, where numerous studies suggest that between 9 and 17 percent of the population were servants. Different estimates of servant populations in Europe can be found in the table 1. The sharp difference between the importance of servants in the examples taken from northern and southern France is striking and suggests strongly divergent practices on this point.[10]

Estimates of southern European servant populations can be found in table 2. In Spain and other southern European societies, even though there were servants both in rural and in urban settings, since the Middle Ages it appears to have been for the most part a job that took young people into households of higher social standing and affected only a small part of the young population in rural areas.[11]

Table 2. Percentage of servants in southern European populations.

	Region	Sample/place	Date	Percent servants
Italy	Kingdom of Naples (south)	Large multi-regional sample	1610-1839	0.7-1.5
	Parma (Po River Valley)	*Contado* (rural areas)	1545	4.0-6.0*
	Pisa (Tuscany)	4 villages, several listings	1656-1740	9.5**
	Bologna	Adjacent rural areas	1853	5.0-7.0*
Portugal	Minho (northwest)	São Tiago de Ronfe (Guimarães) (33 listings, 5 year intervals)	1740-1900	3.6
	Trás-os-Montes (northeast)	Regional rural sample (82 villages)	1796	4.6
	Santarém (central)	Vila de Coruche, Salvaterra de Magos (two villages)	1788, 1789	6.0
Spain	Valencia (east)	Meliana, Benimaclet (2 villages)	1753, 1788	3.8
	Navarre (north)	Large regional sample	1786	7.3
	Santander (north)	Subregional sample (Buelna)	1752	3.0-4.0*
	Galicia (northwest)	Large regional sample	1752	2.6-3.5*
	Basque Country (north)	San Salvardor del Valle, Irún	1766, 1877	5.3
	Cuenca (center)	Large regional sample	1750-1850	3.6-5.0
	Murcia-Alicante (southeast)	Orihuela (Santiago), 4 listings	1719-1829	1.3
	Andalusia (south)	Entire region	1787	2.4

* The data from the following places have been inferred indirectly based on percent households with servants: Parma (10.4 percent of all households with servants; 10.5 percent of the population 15-24 listed as servants), rural areas surrounding Bologna (17.7 percent), and Santander (7.7 percent). For Galicia, estimates based on servants in different social and economic groups. ** These are suburban parishes located only about 2-3 km. from Pisa. This may inform in part the high levels of servants we found.

Sources. Italy: southern Italy (Da Molin; 1990b; also 1990a); Parma (Barbagli, 1984); Pisa (Doveri, 1993); Bologna (Angeli, Bellettini, 1979). Portugal: Minho (Scott, 1998, by kind permission of the author; see also Brandão: 1994); Trás-os-Montes (Mendes, 1995); Santarém (Nazareth, Sousa, 1982; 1983). Spain: Valencia (Garrido Arce, 1992, Pérez García, 1988), Basque Country (Pérez Fuentes, 1993; Urrutikoetxea Lizarraga, 1992); Navarre (Mikelarena, 1995); Galicia (Dubert García, 1992), Santander (Lanza García, 1991), Cuenca (Reher, 1988a; also 1988b); and Murcia (Chacón, 1983); Andalusia (census of Floridablanca).

The census of Floridablanca (1787) in Spain suggests relatively low levels of servants, despite the fact that the data include servants in urban areas where they were frequently more numerous than in the countryside. According to this census, servants made up 22.5 percent of the non-family supply of labor[12] and 2.7 percent of the total population.[13] Despite substantial regional variation, everywhere in *ancien régime* Spain percentages of servants are far below those holding in northern Europe. Much the same appears to hold in the rural areas of Portugal, Italy and perhaps in Greece as well[14], though in cities the importance of domestic servants was normally far higher.[15]

The data taken from numerous local studies before the mid-nineteenth century are corroborated by the first round of modern European censuses. According to the manuscript returns of the census of 1851 in England, servants represented 7.1 percent of the rural population and 3.2 percent of the urban population (Wall, 1983). If lodgers are included in England, these percentages increase to 12.1 percent in rural areas and 14.3 percent in urban areas. In Belgium in 1890, 11.5 percent of the population were servants either in rural or in urban areas (13.7 percent men, 9.6 percent women). In France in 1872 servants represented 6.5 percent of the population (5.2 percent men, 7.6 percent women). In southern Europe the contrasts were very sharp. In Spain according to the 1860 census, 1.3 percent of men and 1.5 percent of women were servants, and in the 1887 census these figures stood at 1.0 and 3.6 percent respectively. In Italy according to the 1861 census, servants represented 2.2 percent of the entire population (1.5 percent of men and 2.9 percent of women).[16]

These data suggest that, despite important local variations, servants were generally between 2 and 4 times more numerous in northern European societies than they were in Mediterranean regions. In the northern part of the continent between 30 and 55 percent of all young persons 15-24 years of age were servants as opposed to southern Europe where between 5 and 20 percent were servants. This means that, on the whole, probably between 50 and 80 percent of young people spent some time of their young lives as servants before marriage in weak-family areas of Europe as opposed to 15-30 percent in the South. In one part of Europe spending a number of years as a live-in servant was the lot of the vast majority of young people, while in another part it was not.

Service had important implications for nuptiality as it was, at least in part, the key to the fairly late marriage age characteristic of the European marriage pattern so aptly described by John Hajnal in his famous article (1965).[17] A close perusal of tables 2 and 3 contained in his article, based on late-nineteenth century census data, reveals that the southern part of Europe did not fully fit the European marriage pattern of late and low intensity nuptiality, though it was fairly far removed from patterns holding in eastern Europe.[18] In Mediterranean Europe, where servants were far less prevalent than in the central and northern part of the continent, nuptiality tended to be somewhat earlier as well. There is now a wealth of research in historical demography which attests to the depth of these differences. While service is not the sole determinant of marriage patterns, it is not a negligible one either.[19]

Despite slightly earlier ages at marriage in Mediterranean Europe, the importance of service as a life cycle activity meant that children ended up leaving home far later in Spain or in Italy than they did in England or in Denmark.[20] For the most part, peasant families in southern Europe with small and medium-sized farms tended to prefer family labor to non-family labor, quite unlike other parts of the continent. In such areas as the southern parts of Spain, Portugal or Italy, where farm size made the exclusive use of family labor impractical, there was an abundant supply of day laborers, who did not co-reside with the farmer and his family.

For most people in southern Europe, then, the real departure of young adults from home only came only with marriage, as opposed to England or Holland where marriage took place after several years away from home and only after young adults had often accumulated substantial savings.[21] In the parts Spain where conjugal families prevailed (center, south and parts of the north), evidence of this practice is abundant, and in diverse historical contexts the percentage of young males heading households in different age groups has been shown to be practically the same as the percentage of married males and the percentage of household heads.[22] Before that time, leaving the parental home was a temporary and even a seasonal phenomenon for most of the young adults. While young girls might go as servants to nearby towns and young men often participated in seasonal migration patterns centered on the harvest or with transhumant livestock, the parental home continued to be the base for most people until the time of marriage. Even in stem-family areas of the Peninsula the situation was similar. For the chosen heir, of course, marriage led to continued co-residence with his parents, but for his siblings leaving home took place only at marriage, those siblings who did not emigrate or enter the clergy basically stayed at home until their marriage. If that marriage proved impossible for one reason or another, they were normally entitled to stay at the family home as long as they wanted or needed. In the few areas of Western Europe where the joint family prevailed, the entire system of household formation was quite different with little or no relationship between marriage and headship.[23]

In England, on the other hand, departure from the parental household took place long before marriage, either as a rural servant or, especially more recently with the decline of agricultural service, as a boarder or lodger in the households of others.[24] In a recent study, Colin Pooley and Jean Turnbull (1997) have estimated that in England between 1850 and 1930, men set up their own households between 2.5 and 5 years before their marriage, and women did so between one and two years before. This situation is a far cry from that holding in Spain, where leaving home before marriage was not only less frequent than in England but also seldom meant that the parental household was abandoned for good. One of the implications of these attitudes was that in northern Europe periods of economic difficulties fell squarely on the shoulders of these young adults, as opposed to the South where they were shared more equally by the entire family group. The protective function of the family in Spain was far greater than it was in England.

There is no evidence whatsoever that these differences between regions of Europe have been erased in recent years. In Spain, for example, the substantial increase since 1977 in the age at which children leave their parental households has been strictly paralleled by the increase in the age at marriage, with both indicators situated today at extremely high levels. In the United States, England, Denmark, and Holland, on the contrary, leaving home long before marriage has tended to be normative behavior. This divergent behavior pattern can be seen in the table 3.

Everywhere, times of economic bounty have tended to bring about younger ages at marriage and earlier emancipation, though leaving home was never as early in Spain as it was in England. In other words, despite fluctuations over time and often regional variability (especially in Spain) in the age at leaving home, areas of strong and of weak family systems always occupy different vital spheres.

Table 3. Percentage of men and women (25-29) still living with parents in different European countries in 1994.

Country	Men	Women
France	22.5	10.3
Germany	28.8	12.7
United Kingdom	20.8	10.8
Spain	64.8	47.6
Greece	62.6	32.1
Italy	66.0	44.1

Source: Fernández Cordón, 1997; Eurostat Labor Force Surveys.

In southern Europe the family takes on many other roles that are largely foreign to its tasks in northern latitudes. Perhaps the most important is the organization of solidarity for the needy and vulnerable in society. The starting point for our discussion of this issue is to consider that vulnerability in historic Europe was sharply constrained by prevailing demographic conditions, especially mortality. Apart from the type of hardship instigated initially and directly by economic factors, it is likely that the incidence of vulnerability in southern Europe always tended to be somewhat higher than in northern Europe because substantially higher levels of adult mortality in those regions led to greater numbers of lone-parent households and to earlier breakups of the marriages of couples past reproductive age.[25] In other words, the "nuclear hardship hypothesis" so aptly described by Peter Laslett would always be more pertinent in high mortality régimes than it was in low mortality ones.[26] Over the past 30 or 40 years, these structural differences caused by mortality have all but disappeared in western European populations.

Traditionally in Mediterranean societies, much of the aid given to the vulnerable members of society came from the family or from individual charity, while in northern societies this was largely accomplished through public and private institutions. The classic example of the institutionalization of solidarity in northern Europe was the English Poor Laws, through which the collectivity came to the aid of the needy and the poor. In Mediterranean Europe the family was essential for the well-being of its more vulnerable members, while elsewhere it was much less so.

Historically the situation of the elderly is a good example of these differences.[27] Before the development of modern pension systems, everywhere a large part of the responsibility for the wellbeing of the elderly fell directly to the family and was based mostly on co-residence with offspring. Despite these similarities, however, in strong-family societies the intervention of the family on this count was much more important than in those societies where weak-family systems prevailed. In Mediterranean Europe, the care of the elderly fell almost exclusively to the family, and was carried out by means of co-residence, the circulation of the elderly among the households of their offspring, or the spatial proximity between the homes of the elderly and those of their children, all of which implied the transfer of goods and services from the families of the offspring toward their elderly parents.[28] In England, on the other hand, the situation was quite different. For one thing, a smaller proportion of the elderly appears to have co-resided with their children.[29] A structural characteristic of English society, epitomized in the Poor Laws, was that the ultimate responsibility for the wellbeing of the elderly fell to the collectivity.[30] In Spain, on the other hand, there were no Poor Laws and only in such cases as extreme poverty or mental or physical illness could people count on institutional support, often organized by the Church. For the vast majority of cases, the family alone took on the responsibility for the material and personal wellbeing of its elderly.

Some years ago I had an informal debate with an English colleague of mine regarding intergenerational family solidarity. I asserted the validity of an idea that was eloquently expressed by an elderly Spanish peasant in an interview several years earlier: "First the children lived off their parents, and later the parents lived off their children. That is just the way life was." My colleague adamantly insisted that this sort of dynamic could never happen in a country like England, where solidarity was expressed through institutions and not through families. It was a society in which children were of relatively little value to their families. We were both correct, each of us based on the societies we knew best.

These differences still exist today. Everywhere, of course, the weight of institutional support has increased with the modernization of society and the increasing longevity of the population, yet divergent patterns of support continue to be visible. In Spain according to the 1991 census, for example, approximately 44 percent of the population older than 60 years of age lived with one of their children.[31] In Nordic countries and in the United States, where co-residence with offspring affects slightly more than 10 percent of the elderly population, the wellbeing of the elderly is based on residential autonomy or on private or public nursing homes which are normally paid for either by public funds, by insurance policies, or directly from the savings of the elderly themselves. In Spain the elderly generally do not have sufficient savings to handle this sort of expense, due in part to the fact that they have been supporting their children for a far greater period of their lives. Despite having comparable proportions of the population in older ages in England and the United States, in southern Europe the number of publicly or privately funded nursing homes for the elderly is minimal compared with other societies. The data in the table 4, containing summary statistics of the importance of institutional living arrangements for the elderly around 1990 in Europe, shows this pattern quite clearly. On this subject, see also De Jong-Gierveld and van Solinge (1995).

Despite recent increases everywhere in the demand for this type of residence, there is no indication whatsoever of a reduction of these divergent residential patterns in contemporary societies. It should be noted that substantial differences in the importance of institutionalized populations also appear in historical data, with far higher levels in northern Europe.[32] In

France during the second half of the eighteenth century, the resources available to formal charitable institutions were far from enough to meet the needs of the poor, be they elderly or not.[33] The situation in much of the rest of Catholic Europe was not unlike the one in France. Where, then, were the poor and the needy in southern Europe who clearly must have been as numerous or more so than in the northern part of the continent? The role of the family in day to day poor relief in southern Europe should not be underestimated.

Table 4. Elder persons living in institutions c.1990 (as % of different age groups).

	Age group	
Country	65+	85+
Belgium	4	-
Denmark	7	-
France	6	20
Germany	5	17
Ireland	5	-
Netherlands	10	47
United Kingdom	5	-
Italy	2	-
Portugal	2	7
Spain	2	6

Source: Jani-Le Bris (1993).

Different attitudes regarding aging and the elderly appear to be rooted in the collective culture of Western Europe. Proof of this is given by the results of a recent survey within the European Union regarding the preferences of the population with respect to the co-residential patterns of the elderly no longer able to live on their own.[34] In Spain, Portugal, Italy, and Greece an average of 74 percent of those surveyed stated that co-residence with children was the preferred option, as opposed to respondents in Denmark, Finland, Sweden, Great Britain, and Holland, where only 25 percent thought the same. The regionalization of these attitudes is not uniform, as shown by the fact that in Scandinavian countries support for co-residence with children is substantially lower than in the United Kingdom, and in some ways the situation of Ireland is more similar to that of Italy than it is to England. Despite this heterogeneity, these data show that the different ways of confronting old age within society exist both in practice and in the popular opinion.

It is instructive to observe that these differences seem to have little to do with the classical types of familial organization existing in Europe, where there were areas of conjugal or nuclear families based on patterns of divisible succession and inheritance (central and southern Italy, Spain and Portugal, central and northern France, a large part of England, etc.), together with others based on the stem family, where designated heirs inherited the bulk of family property on the condition they would continue to co-reside with their parents after marriage (much of central Europe and Scandinavia, Scotland, part of the Low Countries, much of northern Spain and Portugal, and the mountainous regions of the Pyrenees, the Alps, and the Massif Central)[35]. Indeed the strong families we have seen flourished not only in areas of stem families, much as weak family systems were not restricted to conjugal family regions.[36]

Historically the strength of familial ties appears to have conditioned the way in which succession was carried out in stem-family regions. In Catalonia and the Basque Country of Spain, stem-family areas *par excellence*, the obligation to co-reside *en una mesa y compañía* ('at one table and in the company of') with the parents was normally stipulated quite simply in the marriage contract (*capitulaciones matrimoniales*). In much of central and northern Europe, veritable retirement contracts between parents and their children were drawn up listing in great detail the rights and obligations of children and parents.[37] These contracts, which originally did not necessarily even involve kin, were designed to safeguard the wellbeing of the elder generation and to facilitate the emergence of inheritance *intervivos* (Gaunt, 1983). The history of these contracts was frequently fraught with the intergenerational strife they were designed to minimize.[38] Contracts such as these are simply unimaginable in a southern European context. In other words, succession itself within stem-family systems appears to have been conditioned by the strength of familial loyalties and solidarities holding in any given region of Europe.

In Spain outside of the stem-family regions, parents also facilitated a type of inheritance *intervivos* for their children, but these arrangements were invariably informal and seldom contained terms stipulating how intergenerational support mechanisms were to be implemented (Reher, 1997).[39] Of course, family strife has existed in every culture. What is instructive in this comparison is the fact that formal and informal retirement arrangements in Spain, and likely throughout southern Europe, had little to say regarding the day to day dealings between parents and their children, as opposed to northern Europe where the key to an amicable arrangement was that it was thorough to the very minute details. In Mediterranean Europe these arrangements were done within the context of a culture where strong family ties were an essential component and intergenerational relationships were strictly and normatively controlled, as opposed to Germanic Europe where this cultural component was not nearly as strong.

Each of these family systems has ended up generating justifications that are coherent with its own premises. In weak-family areas, the value attributed to the individual and to individualism tends to predominate.[40] Young adults leave home, encouraged by their parents, so as to acquire the experiences they need to handle life as autonomous individuals. Leaving home at an early age is considered an important part of their education. Where the strong family flourishes, the familial group more than the individual tends to predominate in the socialization of the young. In these contexts, the family is seen as defending its members against the difficulties imposed by social and economic realities. A child receives support and protection until he leaves home for good, normally for marriage, and even later.

Faced with the transition to old age, in one context individuals attempt to prolong their physical independence as long as possible and, when this is no longer feasible, to conserve a measure of economic independence that will enable them to enter a rest home or afford some other solution. They would never give serious consideration to going to live with their children; nor would it enter the minds of their children to have their elderly parents at home with them. This attitude is so widely held in the United States, for example, that the elderly who do live with their children probably tend to come from strong-family ethnic backgrounds.[41] In sharp contrast to this pattern, in areas of strong families, maintaining independence as a matter of principle would seem like nonsense, and this only happens when, for one reason or another, there is no family. In Spain it has always been said that the only truly poor person is the person who has no family. Furthermore, the solidarity between the older and the younger generation never breaks down; it is a social obligation expected by individuals and by their families. The elderly who do not maintain regular contact with their children are a small minority of the population, much as are the aged in weak-family societies who receive regular weekly or daily visits from their children. In both situations there is intergenerational reciprocity, though it is always understood quite differently. These are divergent modes of behavior, applied in each context with a maximum of good will.

3. UNCERTAIN BUT DISTANT ORIGINS

The social, economic, and even demographic explanations normally used to explain the origins of these diverse ways of living family life are not convincing, though their geography is fairly clear: the Mediterranean region has strong families, while the northern part of the continent is characterized by their weak families. In between, nations like France and, to a lesser extent, Germany do not fit easily into either system, and constitute a good indication that our portrayal simplifies in the interests of clarity a very heterogeneous European experience. The differences we have pointed out are visible as early as we have empirical data to test for their existence (the seventeenth century, more or less), though it seems likely that they were in place long before.

The basic geography of our family forms suggests that their origin is related with Roman and Germanic-Nordic Europe, and may well have been forged at least initially during the later part of the Roman Empire and the early Middle Ages. In a thought-provoking book on the family and marriage in Europe, Jack Goody (1983) traces the roots of the western family tradition back to the fourth century. Before that, in the ancient world on both sides of the Mediterranean and in the Near East there were certain common characteristics of familial organization among which he emphasizes the existence of patrilineal clans, the ability of both men and women to inherit and, perhaps more important, the fact that most marriage was strictly endogamous, the kin group had great importance and the conjugal family did not, and in social and cultural terms women had relatively little importance.[42] This 'oriental' family system was replaced during the late Roman Empire by an 'occidental' structure in which the basic cell of social organization became the conjugal pair and norms for marriage outside the kin group were strictly enforced. Goody feels that this change, which led to a fundamental weakening in the ties of the kin group in favor of the primacy of conjugal marriage, began during the late Empire, gradually became a structural characteristic of the entire Christian world over the next millennium, and formed the basis for a family system in the West which gave rise, among other things, to its characteristic marriage pattern (Hajnal, 1965).[43]

Here we would like to suggest that perhaps the implantation of this western family structure in Europe was not uniform. In the northern part of the continent, Christianized forms of familial organization ended up meshing gradually with existing Germanic legal and social traditions based, among other things, on the importance of the tribe, the individual, and the relevant social position of women.[44] In southern Europe the influence of the Germanic tribes was much more superficial and short-lived. Besides, from the early eighth century on there were a series of Muslim incursions, strongest in Spain and in the Balkan Peninsula but also present in southern Italy, which tended to bring back, once again, oriental family structures, so central to Islamic societies, that are based on the overriding importance of kin ties.[45] Even where the Muslim occupation was short-lived, the geographical proximity of oriental family systems in North Africa could not help but influence the development of the family in southern Europe. What arose in those areas was a family system which in all likelihood was hybrid in nature with a basic western structure but also with certain oriental trappings centered especially on the importance of kin ties and extended family loyalties.[46]

The Reformation, with its emphasis on the individual and self-reliance, on the value of work, on a this-worldly asceticism, and on predestination proved a sharp contrast to Catholicism, based on authority, the other-worldly and spirituality. These contrasts had fundamental implications for family life and for the economic and social organization of European society. For Protestant reformers, marriage ceased being a sacrament and became a civil contract governed by matrimonial tribunals, and many of the traditional Catholic constraints on marriage (e.g. consanguineous marriages to the seventh degree) were either reformed or repealed.[47] More important, perhaps, is that the home itself became a place of self-fulfillment and of sharing. Most notably in Calvinism an emphasis was placed on marriage not so much as a context for reproduction, but rather as a partnership in the garden of the Lord, in the rearing of children in the faith, and the advancement of God's Kingdom. In so doing, the Reformation had laid the grounds for marriage as a partnership, so essential for northern European marriage systems and for the full development of potential of individuals in this world, all in sharp contrast to the Catholic world where parental authority and family loyalties tended to be far more hierarchically structured.[48] By implication, the Reformation ended up enhancing women's position in society as opposed to the Europe of the Counter-Reformation where, despite luminous figures like St. Teresa of Avila and others, the position of women seems to have undergone little change before the eighteenth century or even later.[49]

These fundamentally different attitudes towards life and religion settled on a continent where divergences in family systems had already been developing for over a 1000 years. It can be argued that the progress of the Reformation was itself facilitated and influenced by the differing attitudes towards the individual and family life existing in Europe during the medieval and early-modern periods. It is also unquestionable, however, that the Reformation tended to deepen and solidify the age-old north-south contrasts in Europe. It is interesting to note that in Catholic nations of northern Europe forms of familial organization tend to diverge at least partially from the prevailing patterns. Ireland is a excellent example of this: a decidedly Catholic nation in northern Europe whose external family forms fit quite poorly with our north-south comparisons.

The effects of the Industrial Revolution were felt first and most profoundly in northern Europe and this can be related both as cause and as consequence of its prevailing family system. John Hajnal (1983) has discussed this process from one vantage point, arguing that family and marriage patterns contributed to the low-pressure demographic regimes existing there and ultimately to the flowering of northern European economic growth after the second half of the eighteenth century.[50] The Industrial Revolution, based an ethic in which the economic rationality and creativity of individuals[51] was paramount, ended up reinforcing an individual-oriented family system in the industrializing areas well before this same process began to take effect in most of the southern part of the continent. Thus, at least in its origin, the entire process of economic modernization would seem to have reinforced the prevailing family differences in Europe.

Regardless of their historical origins, attitudes toward the family and the individual make up the cultural tapestry of societies, and thus they are models that are learned at very young ages and that societies –individuals, families, institutions– help perpetuate. Learning these behavior patterns is the cornerstone of the socialization of children. They are attitudes shared by the society as a whole. Perhaps because of this they have been so resistant to the otherwise corrosive effects of economic, political, social and demographic modernization. It is undeniable that the changes of this past century have tended to make cultures and mentalities more uniform, though they have done little to erase the historic profiles family systems in Europe.

We have described family systems in which either the individual takes precedence over the family group or the individual develops his personality and even his freedom within the family group. The systems we have described are by no means the only ones possible, though they are the most widespread in western Europe. Had we wanted to enlarge our perspective on this point, it would have been necessary to consider the type of family found, for example, in eastern Europe or in areas of the Muslim world and Asia, especially China, where the weight of the extended kin group is far greater than it is in southern Europe. At this level, our strong family from Mediterranean Europe in reality lies somewhere between the individualism characteristic of northern Europe and North America and the strict allegiances and corporatism generated within enlarged families lineages, and clans that have characterized large regions of Asia.[52]

4. SOME IMPLICATIONS OF FAMILY SYSTEMS FOR SOCIETY

Family systems are neither good nor bad, but they are not neutral either. They do much to characterize the societies that envelope them. Many of the differences distinguishing European societies are derived directly or indirectly from the nature of their prevailing family systems. The Church and the State have been aware of this for centuries, and here we are not saying anything surprising. It might prove instructive, however, to point out some of these differences existing in the West at the end of the millennium, because that may enable us to rediscover the importance of the family, an institution apparently given up for dead by many students of contemporary society.

Table 5. Different social indicators related to family behavior in several European countries.

Country	Crude divorce rate, 1995	Non-married couples cohabitating, 30-44 (% of total population in age group), 1993	Births outside marriage (% of all live births), 1995
Austria	2.3	-	27.4
Belgium	3.5	9.2	15.0
Denmark	2.5	20.9	46.5
Germany	2.1	5.9	16.1
France	2.0	11.5	37.2
Ireland	-	1.7	22.7
Luxembourg	1.8	7.7	13.1
Netherlands	2.2	-	15.5
Finland	2.7	-	33.1
Sweden	2.6	-	53.0
UK	2.9	9.3	33.6
Greece	1.1	1.0	3.0
Spain	0.8	2.9	10.8
Italy	0.5	2.2	8.1
Portugal	1.2	1.4	18.7

Source: European Observatory on National Family Policies, 1998; Statistiques en Bref, 1997.

Societies with strong families tend to have greater social cohesion. The low incidence of divorce and extramarital pregnancy in them is a good example of this. Strong-family societies are usually more conservative than weak-family ones in social -though not necessarily in political- terms. In other words, the social control of behavior tends to be more effective in strong-family societies. The majority of the social indicators related directly or indirectly to the family seem to indicate this. The indicators in table 5 are good examples of these differences.

The case of Portugal, with a very high incidence of births outside marriage, has always been exceptional in southern Europe and is based largely on the fact that for well over two centuries strong male emigration has left a society with far fewer men than women. Some people have attempted to explain these social differences in terms of religious attitudes or by the position each society occupies on its particular road to modernization. They can be more easily explained, however, by the nature of the family systems prevailing in particular societies. The problem of the homeless is a prototypical example of the lack of social cohesion in contemporary society. It is often surprising to note that the incidence of the homeless is much greater in the United States, for example, than it is in Italy, Spain, or Portugal, despite the greater economic dynamism and higher living standards and the lower levels of unemployment of the United States.

What has happened? In all probability, families in Mediterranean Europe have absorbed a part of this mass of uprooted people who in northern Europe and the United States have had to fend for themselves, either on the public dole or with private charity.

The subject of unemployment is intriguing on this count. An apparent contradiction in Spain, for example, is that it has extremely high levels of unemployment, yet people seem to live modestly well and the external indicators of social distress are fairly muted, at least in comparison with other countries where unemployment is far lower. In Spain, the essential mechanisms of familial solidarity stipulate that the family group protect its members from the vagaries of employment and economic cycles, and thus the social implications of unemployment tend to be hidden, at least in part, within the family. If a country like the United States, for example, had similar levels of unemployment, uprootedness would have been widespread and the social and political consequences enormous. In Spain, comparatively little social disruption has taken place, largely because of the role of the family.

Another eloquent example of how southern European families actively intervene in the wellbeing of their own members can be seen in lone-parent households. Due to divorce and teenage pregnancies, everywhere in recent years the number of these types of households has been on the rise, though there continue to be important and now familiar north/south differences in levels.[53]

Two recent studies of this phenomenon in Spain have brought to light the existence of a very significant 'grandmother effect', whereby high percentages of single mothers end up co-residing in the same household as their own mothers (the grandmother).[54] In 1991 nearly 30 percent of all lone mothers (independent of their marital status) with children under 18 co-resided with their own mothers, as did slightly under half of all mothers with children under 6.[55] Interviews have confirmed that even when direct co-residence was not the case, grandmothers often lived nearby and were essential in helping the mothers care for the children and secure a job to support their families. In more general terms, the intervention of grandparents is not restricted to exceptional situations such as lone-parenthood, but is a structural characteristic of family life in Spain. The massive entry of women into the labor market in recent years in Spain has been facilitated to a large degree by the existence of grandparents willing and able to help care for the children when their parents are at work.[56] Though the importance of the 'grandmother' effect has been noted in very diverse contexts, it would seem unlikely that this sort of straightforward type of familial solidarity has ever been as important in northern Europe or the United States as it is in the Mediterranean region.[57]

Loneliness is one of the most important social problems in weak-family societies. I refer to the loneliness of the individual who must confront the world and his own life without the safety net of familial support so characteristic of strong-family regions. A proxy for loneliness in society is the proportion of single-person households (table 6). Here once again the differences between northern and southern Europe on this count is immense.

Suicide, an indirect consequence of loneliness, tends to be far higher in northern Europe and the United States than it is in southern Europe, as data in table 7 clearly show. The effects of loneliness are compensated in weak-family societies by means of a strong tradition of civic association, where people form groups, clubs, and societies for the most varied purposes. The number and variety of these associations in England or the United States would be unimaginable for a citizen of southern Europe. In weak-family societies the individual is able to combat loneliness by turning directly to civil society, itself largely the product of the needs and initiatives of its members, in contrast to strong-family societies where the family comes between the individual and civil society, covering a large part of the needs stemming from loneliness.[58]

Table 6. Percentage of single-person households in Europe, 1990/1991.

Country	Percent
Austria	32.3
Belgium	31.4
Denmark	38.1
Finland	38.0
France	29.2
Germany	37.7
Ireland	27.1
Netherlands	37.7
Sweden	44.0
United Kingdom	30.0
Greece	21.1
Italy	23.7
Portugal	16.6
Spain	16.9
European Union	30.3

Source: Statistiques en Bref. Population et conditions sociales (1995/5).

Weak-family societies, then, tend to be associational societies with a deep civil component, and strong-family ones tend to be more passive societies, at least in terms of the importance of individual initiatives within them. The sense of individual responsibility for collective norms and needs, so essential for our concept of democracy and civil society in the West, is often conspicuously absent from southern European societies, while in northern societies it is an integral part of the social fabric. In sum, the countries of northern Europe and of North America have well-developed civil societies that thrive on individual initiatives, but with a dark side shown by their lack of social cohesion and by the desperation and anguish so prevalent in them. They are tough societies, but they are also dynamic ones. Mediterranean societies are more pleasant, more comfortable, more conformist, more oriented towards the family group, and less dynamic.

The increasingly rapid process of aging affecting contemporary society is one of the most important challenges confronting developed societies today. Meeting this challenge will not be the same in societies where strong-family systems prevail as in those with weak-family systems. Strong-family societies can and should count on the institution of the family when planning means of support for the elderly. This support can be expected to begin when the health of the elderly is still good, and will be offered by means of co-residence or residential proximity with frequent personal contact. Once health begins to deteriorate, the family will continue to be essential both as a place for the elderly to live and as a source of company for them. This last aspect is difficult to measure empirically though it makes up an important part of the wellbeing of the elderly. It is also likely that strong families will continue to be a source of income supplement for their needy elderly, much as they have done until now.

Table 7. Deaths by suicide in Europe, 1990 (per 100,000).

Country	Men	Women
Belgium	25.0	9.4
Denmark	30.7	14.8
Finland	47.9	11.8
France	29.1	10.2
Germany	23.6	8.6
Ireland	16.1	5.3
Netherlands	12.1	6.8
Sweden	22.5	9.7
United Kingdom	12.2	3.6
Italy	10.7	3.6
Portugal	13.9	4.1
Spain	11.2	3.6
European Union	18.9	6.6

Source: Eurostat, Yearbook, 1997.

In weak-family areas, the care of the elderly will be based much more on individual savings, on residential autonomy, on retirement communities and nursing homes, and on the support of public institutions. The most common pattern of behavior will be for an elderly person to move directly from an autonomous residence to a nursing home, normally obviating the need for an intermediate stage of co-residence with a child, so frequent in southern Europe. Social and emotional support for the elderly will be offered by charitable institutions and volunteers, as well as by the families themselves, though on this point a large number of families will not be up to the task. Institutional care is much more costly and demanding for society than care based on the family. Yet in relative terms the level of savings among the elderly will tend to be greater in weak-family areas, enabling many elderly to contribute economically to the costs of their own care. Implicit in this same context is the fact that economic abuse of the elderly, a crime on the rise the world over, will always be more common where personal savings are greater and where there is less family influence on those savings. Family-based abuse, on the other hand, may well be more prevalent in strong-family societies, though here the active participation of the entire family in the welfare of the elderly will minimize the ability of certain individuals to manipulate elderly parents to their own ends.

Politicians, government officials, and public planners would do well to bear in mind the specific characteristics of family systems when designing social policies affecting the elderly, because the effectiveness and success of these policies will depend on how well-tuned they are to these characteristics. Everywhere, of course, promoting healthy living among the elderly as well as keeping pension systems afloat will have priority, though specific policies will work differently in different family systems. In strong-family areas, for example, the protection of the elderly should include support for the family in carrying out its traditional role of attending to the needs of the elderly. Where weak-family systems prevail, on the other hand, stimulating individual savings and the work of charitable groups as well as safeguarding the elderly from predators will all be essential.[59]

It is evident that the nature of family systems and their loyalties does not fully explain these social differences, though understanding them adequately is impossible without keeping in mind the importance of the family. It is within the family that the way in which the individual relates to the family group and to society is first learned. This learning process is deep and lasting, and during the rest of our lives we end up implementing the behavioral norms we learned during our infancy. They are norms that life itself ends up confirming all the time.

5. PRESENT AND FUTURE CHALLENGES TO FAMILY SYSTEMS

It would be incorrect to believe that familial forms are unchanging. Throughout history the family has been changing, and it continues to do so today. Unquestionably one of the main destabilizing factors in the contemporary world is the new demographic regime that affects all of us. This demographic challenge has two principal characteristics. For one, there has been an important reduction in mortality, especially among adults and the elderly, thus leading to increasing numbers of elderly persons who are spending increasing numbers of years in that stage of their lives. Though the scientific evidence is still inadequate on this issue, it is also possible that the elderly will end up spending an increasing proportion of their lives with precarious mental or physical health, thus making them still more vulnerable. The second characteristic is that in the past 20-30 years there has also been a drastic reduction in fertility in most western societies, with indicators currently at the lowest levels ever attained and with rapidly declining numbers of births. This demographic regime has produced an extremely rapid process of aging, with the elderly occupying ever greater proportions of the population.

Perhaps more pertinent for the subject at hand is the fact that everywhere families find themselves with ever-fewer children and ever-more elderly members. The demographic balance of the family group has been broken and is now in rapid transformation. While differences do exist, in most developed societies the demographic context is basically the same. This reality is vital for the family, and its consequences will likely be far greater in strong-family areas than in those where the family tends to be weak. Where strong families prevail, the support children for their elderly parents is closely dependent on the fact that there are enough children to take care of their parents. Yet recent demographic change has altered this circumstance and there is now a possibility that the family group in southern Europe will end up having as many dependent as active members.[60] In weak-family societies, this challenge will tend to be posed in terms of the society as a whole, and somewhat less so in terms of the family, mainly because the type of familial solidarity so characteristic elsewhere is much less decisive within society. As a result, strong-family systems appear to be much more vulnerable to the effects of demographic change than do weak-family systems.

We might wonder whether the differences described here will continue to characterize European societies or whether there will be some sort of convergence in family forms will occur. Recently Louis Roussel (1992) proposed a model for the future development of the family in western Europe. For him, there is a process of convergence afoot on the continent that eventually will render the family similar in Germany and in France, in Sweden and in Spain. He feels that in the more 'advanced' northern nations, the rates of change will slow, while they will continue to be high in the southern flank of Europe. The end result will be a truly 'European' family for the first time. Roussel's idea is attractive, especially because it emphasizes the commonality of European experience.

Nevertheless, I cannot but disagree with this idea, mainly because its underpinnings appear to be antihistorical. At the very least they tend to minimize the depth of cultural and historical differences in Europe. Once again it is as though modern society had finally done away with the pernicious effects of history, launching us toward the adventure of the future. It is a type of neo-modernization discourse in which economic and social change torches all vestiges of cultural and historical difference. This seems hardly likely because these are differences have characterized European societies for centuries, and it would not be prudent write their death certificate too hastily.

It is unquestionable that in Europe certain external indicators of the family and of family forms are converging: the importance of solitary households is increasing, the weight of extended families is decreasing, fertility and nuptiality are declining, and the number of children born out of wedlock is on the rise. Additionally, parental authority has diminished, improvements in health and social welfare have led the elderly to maintain their own independence much longer before going to the State or to the family for help, children and women have acquired far greater autonomy with respect to the familial group, and women have entered the labor market in great numbers. It is also true that the rate of change in much of southern Europe over the past 15-20 years has often been dramatic. These are all indisputable signs of the times and they affect all western societies.

But does this mean that European families are on the path to uniformity, much as Roussel seems to suggest?[61] Perhaps not. For one thing, despite general moves in the same direction, most of these indicators show no decline in relative variability whatsoever. Even with the great transformations in recent years, the rank order of European countries in most instances has remained unchanged. Perhaps more important, we should recall that the family is an institution that is far more complex than we might suspect when using straightforward empirical indicators reflecting certain types of behavior bearing on the family. People's attitudes toward the family, the way they live family life, and the type of influence the family has over the lives of its members are essential to the meaning of the family; and there is no indication whatsoever of convergence on this count.

My guess is that the outcome of these changes will be a convergence in the external indicators of family life, but this convergence will not undermine the deep disparities that have always characterized the family in the different regions and cultures of Europe. The forces making up the contemporary world, common for the most part in all societies, are not the only factors forming these societies, because societies' own historical trajectories, different in each case, will also contribute to the specific contours of the present and the future. This concept, known as 'path dependency,' refers to a simple but important reality. No matter how nearly universal the factors of modernization may be, once they enter into contact with different historical, cultural, geographical, or social realities, the end result will necessarily be different in each context. The confluence of factors of change and of structural realities, with different results every time, has occurred many times in the past, and there is no reason that the near future should be any different. It is worth invoking this concept here because it underscores the fact that the realities of the present-day world can never be adequately understood without bearing in mind both contemporary forces and historical traditions.

In the future the Spanish family will continue to be traditional and strong, the English family traditional and weak. Spaniards and Italians will continue to care for their ill and vulnerable parents, just as grandparents take care of their children's offspring when they are at work. The English, the Americans, and the Swedes will continue to maintain their commitments to individualism and to residential autonomy. Spaniards will continue to remain at home until they get their first stable job, and Nordic adolescents will continue to actively seek their liberation from the family ties that bind them. Demographic change will have more severe effects in the South, making the State and personal savings play a greater role in the wellbeing of the elderly, though this will not replace the role of the family in a significant way. The future promises to bring many changes, but weak-family and strong-family systems will continue to occupy clearly differentiated vital spheres. Appreciating the strength, flexibility, and resilience of the modern family continues to be essential for any viable understanding of society.

David S. Reher, Department of Sociology, Human Ecology and Population, University of Madrid.

NOTES

[1] Scandinavia includes Iceland but not Finland. Our definition of northern Europe for the purposes of this paper is very similar to the one used by Hajnal (1982) for northwest Europe.

[2] Here we are referring mainly to Portugal, Spain and Italy, though at times in this text southern France and Greece are included as well.

[3] Both John Hajnal (1965, 1982) and Jack Goody (1996a) have pointed to these two factors as defining traits of what has been called the 'Northwest European household formation system.' These authors have concentrated more on comparing these patterns in historical contexts with eastern European or non-European populations than on detailing north/south differences within Europe. See, for example, Hajnal (1982).

[4] Approximately translated: "The bride (or groom) demands a home."

[5] On this point see, for example, Reher (1998).

[6] An example of this type of behavior can be found in the town of Cuenca (Spain) during the nineteenth century, where more than half of all newlyweds lived for some time in either the bride's or the groom's family household. This type of coresidence, however, was always temporary. See Reher (1990).

[7] There are indications that the institution of agricultural service in England dates at least as far back as the 1377 Poll Tax where one-third of all farmers had servants (Smith, 1981).

[8] These percentages varied widely by locality. For more on the importance of agricultural service as a life-cycle activity, see, for example, Laslett (1977a; 1977b) and Wall (1983a).

[9] For a sample of 21 English communities, Richard Wall estimated that between 10-14 and 15-19 years of age, almost half of the people had already left their family homes and had entered agricultural service in other households or were lodgers in autonomous residences. See Wall (1987; also 1978).

[10] Mitterauer and Sieder (1977) have estimated that on average between 7 and 15 percent of the population in preindustrial times were servants in northern Europe. For slightly higher estimates, see Burguière (1986).

[11] In his comments on the study of Herlihy and Kaplisch-Zuber (1978), Richard Smith (1981) points out the relative absence of servants in Tuscany (only 0.2 percent) based on the Florentine *Catasto*, as opposed to the bountiful presence of servants in England during the same period. For more on servants in the 1427 *Catasto*, see Klapisch (1972).

[12] Here we consider 'non-family labor' to be the sum of servants (*criados*) plus day laborers (*jornaleros*).

[13] Ten years later, the census of Godoy (1797) showed that servants made up 1.7 percent of the population of Spain, and 17.8 percent of the non-family supply of labor.

[14] Regional differences in southern Europe were considerable, as has come to light in several local studies, with northern regions showing higher proportions of servants than elsewhere (e.g. for Spain, see Reher *et al.*, 1993). Nevertheless, levels of servants were never as high as they were in England or in other parts of central and northern Europe. In Greece, studies suggest that agricultural service was relatively unimportant and that emancipation from parental households for most people came only with marriage. On this point, see Osswald (1990) and Hionidou (1995).

[15] Everywhere in southern Europe cities had far higher levels of servants than rural areas. In the city of Parma in 1545, for example, 30.7 percent of all households had servants as opposed to only 10.4 percent in the *Contado*. For data on Parma and on other sixteenth century Italian cities, see Barbagli (1984). See also Arru, 1990). In the small town of Cuenca (Spain) in 1800, for example, servants made up 11.0 percent of the total population and 25.4 percent of households had servants, as opposed to rural areas where servants represented only between 3.6 and 5 percent of the total population during that period (Reher, 1990, 205). According to the census of Floridablanca (1787) servants represented 11.7 percent of the population of the city of Madrid, as opposed to only 2.7 percent of the population of Spain.

[16] We have been unable to make use of censuses from other European countries because the occupational structure used in them makes identifying servants in rural areas extremely problematic.

[17] For the importance of service for late marriage age, see also Hajnal (1982).

[18] This can also be seen quite vividly in the map of I_m (female nuptiality) around 1870 based on the data compiled during the Princeton European Fertitlity Project (Coale and Watkins, 1986).

[19] Hajnal (1965) and other authors have made the point that later marriage was linked to higher standards of living.

[20] As opposed to the 50 percent of the population aged 10 to 19 living away from home in England (Wall, 1987), Reher (1988) found that in rural areas of Cuenca during the nineteenth century around 90 percent of people of the same age groups continued to reside in the households of their parents.

[21] According to McIntosh (1984), for example, during the second half of the sixteenth century in Essex, where servants made up 20 percent of the total population, young adults left their familial homes as adolescents and spent between five and ten years as servants before setting up a home of their own.

[22] On this point, see, for example, Reher (1997).

[23] In Western Europe, joint-family systems were only found in areas of central Italy and in certain parts of central France between the Franche-Comté and the Massif-Central (Burguière, 1997; also 1986). In joint family systems the gap between proportions of ever-married men and proportions of household heads or married household heads would be much greater than in stem-family systems. The classic example of this can be found in Tuscany after the Black Death (1427-1430) where the highest proportions of ever-married men is reached around 40 years of age while peak levels of headship are reached much later in life. The situation of medieval Tuscany, with a prevalence of joint family households, late male age at marriage together with very young female age at marriage, is quite exceptional in Europe. See Hajnal (1983) and Klapisch and Demonet (1972).

[24] The importance of boarders might be significant in Northern Europe and in American society. This is pertinent to our argument when the lodgers were young, as they often were (Wall, 1983b). Numerous studies covering England suggest that lodgers represented between 4.9 and 5.8 percent of the population of rural England between 1650 and 1821 (Wall, 1983a) Their presence in industrial areas was always greater than in the countryside. In the census of 1851 of England, lodgers made up 11.2 percent of the urban population of England as opposed to 5.0 percent in rural areas (Wall, op.cit.). See also, Anderson (1972), Glasco (1977), Modell and Hareven (1977), and Blumin (1977). Under certain exceptional economic circumstances, lodgers might also be very important in nineteenth century Spain. An example of this is San Salvador del Valle, a mining settlement not far from Bilbao in northern Spain where lodgers made up between 25 and 30 percent of the total population between 1887 and 1900. See Pérez-Fuentes (1993).

[25] During the seventeenth and eighteenth centuries, life expectancy at birth (e_0) in southern European countries was between 25 and 30 years, as opposed to northern Europe where it varied between about 33 and 40 years.

[26] For more on this hypothesis, see Laslett (1988). For estimates on the incidence of vulnerability over the life-course in a pre-industrial Spanish population, see Reher (1997).

[27] For a discussion of systems of support for the elderly during past times, within the context of a basic Northwest Europe – non European comparison, see Hajnal (1982) and Goody (1996a).

[28] For more on the circulation of the elderly among the households of their children, often called *ir por meses* in Spain, see Reher (1988). An example of family groups maintaining patterns of residential proximity, even in urban contexts and over the course of several generations, can be found in the case of the Recuenco family in the town of Cuenca during the eighteenth and nineteenth centuries (Reher, 1990).

[29] Based on our still inadequate understanding of the co-residence patterns of the elderly, it appears that between 40 and 50 percent of the English elderly co-resided with their children during the eighteenth century, as opposed to somewhat higher values in Spain at later dates. For the English data, see Laslett (1977; 1989) and Wall (1984; 1995). In the northern Spanish city of Bilbao between 1825 and 1935 approximately 70 percent of the ever-married elderly continued to live with their children (Pérez-Fuentes and Pareja, 1997).

[30] On this point, see R. Smith (1984) who questions the idea that the elderly depended exclusively on the family, emphasizing the importance of the structural dependence of the elderly on the collectivity, especially in England. According to Smith "...From a very early period in English history, and in other Northwestern European areas, it seems that 'risk devolution' and poor relief have been centered on the community rather than on the family." One of the reasons for this was that the needs of the elderly were greatest just when the households of their offspring were undergoing particularly difficult economic times due to the presence of young children at home (R. Smith, 1984; Anderson, 1977). For more on the role of the collectivity in the support of the widowed elderly, see J. Smith (1984), Laslett (1984; 1988; 1989). J. Smith (1984) underscores the inability of households to generate additional income in order to maintain their economically inactive elderly, thus making the flow of income from outside the household totally necessary. David Thomson (1984) has suggested that the benefits of social welfare going to the elderly British today are somewhat lower in relative terms than the pensions paid during the first half of the twentieth century, and much lower than the income transfers derived from the Poor Laws during the nineteenth century. Elsewhere Thomson (1991) has gone so far as to affirm "It is unEnglish behaviour to expect children to support their parents." In a recent paper, Pat Thane (1998) has argued that both in the recent and more distant past of England the family has played a complementary role to that of the community in supporting the elderly. For a partially divergent point of view on this issue, see Kertzer (1995).

[31] This percent varies by age, with fairly high levels for persons aged 60-69 due to families with children still at home (45-50 percent), somewhat lower levels for persons 70-79 (33-35 percent), and then much higher levels for persons over 80 years of age (>50 percent).

[32] Richard Wall (1984) has found that around 5 percent of the population above 60 lived in institutions in several English communities during the eighteenth century. In Spain, both according to the census of Floridablanca (1787) and that of Godoy

(1797), somewhat below 1.4 percent of the population above 50 resided in institutions. This last percentage is based on the supposition that all the physically ill, mentally deranged or indigent people residing in hospitals or charitable institutions (*Casas de Miserecordia*) were above 50 years of age. It should also be kept in mind that the Spanish data include urban areas where the weight of institutions was far greater than in the rural world. In other words, our estimation procedure tends to overestimate the importance of the institutionalized elderly present on these censuses, and thus the comparison with English figures tends to understate the differences.

[33] In France the system of poor relief was similar for the most part with that holding in much of Mediterranean Europe and was based on private almsgiving and donations, and on institutions erected in the spirit of the Counter-Reformation. Olwen Hufton has estimated that the total resources available to these charitable institutions would not have been enough in any one year to buy a single pound of bread for each hungry person. See Hufton (1974). It is also interesting to note, however, that in 1791 the degree of institutional help was considerably greater in northern France than in the southern part of the country (Hufton, 1974).

[34] The results of this survey are reported in van Nimwegan and Moors (1997).

[35] For more on the geography of these family forms, see Todd (1990).

[36] For an attempt to rethink the implications for people's lives of these classic family systems existing in Europe, see Kertzer (1989; 1995).

[37] These agreements might stipulate, for example, whether or not the parents could sit next to the fireplace, what they could eat, or other seemingly minute aspects of daily life. The use of these agreements was widespread and had existed since the Middle Ages in areas of Europe where Germanic Law had prevailed (Scandinavia, Germany, Austria, England, Bohemia, Moravia and Finland) (Mitterauer and Sieder, 1977; Gaunt, 1983). For an example of this type of contract between father and son, see Ehmer (1998) and Gaunt (1983).

[38] In 1772 the agricultural reformer and traveler Anders Bachaeus reported from one central Swedish parish that the young went about calling the retired 'the old devils' and demanding, "What is their purpose in living?" (cited in Gaunt, 1983).

[39] In Cuenca and probably in much of central and southern Spain, only about 10-20 percent of property owners even bothered to draw up a will (Reher, 1988a).

[40] For more on this subject, see MacFarlane (1978).

[41] In historical contexts differential behavior patterns regarding the family have appeared in different ethnic groups. In her study of the family and the elderly in the State of New York during the 1920s, for example, Weiler (1986) found that: "The immigrants from eastern and southern Europe stressed the value of children as insurance in old age, whereas Americans and west Europeans valued individualism and independence between generations." See also Chudacoff and Hareven (1979). Regarding more general aspects of familial organization, cultural contrasts have appeared in studies such as those of Carroll (1988) and Glasco (1977).

[42] For a more complete portrayal of western and eastern social structures, see Goody (1983) and Guichard (1977) whose ideas are the starting point for Goody's essay.

[43] Goody attributes the beginnings of this change to the Christianization of the late Empire and in particular to a Church policy designed to undermine the traditional clan and kin networks for its own material benefit (Goody, 1983).
[44] Goody affirms that the early Christian missionaries in northern Europe went to great lengths to change a number of the Germanic practices centering on strategies an individual might adopt in order to produce an heir which were more like those of earlier Mediterranean cultures (Goody, 1983). Nevertheless, Tacitus (*Germania*) remarked on a number of aspects of Germanic social organization and behavior which suggest the existence of quite 'western' types of familial organization even in pre-Christian times, as well as an emphasis on the importance of independence and individuality. He suggests that marriage was later for men and women (chapter 19) and that it was viewed as a shared pact ("...she is thus warned by the very rites with which her marriage begins that she comes to share hard work and peril...") to be used in work, war and reproduction (husbands brought gifts to the marriage which included tools, animals and weapons, and the wives contributed a piece of armour) (chp. 18). There were injunctions against infanticide (19) and mothers practiced breastfeeding of their infants (20). Girls and boys were educated in the same way (20), and people felt the need to live in widely scattered houses with plenty of space around them (16). Houses were set up "...according as spring-water, meadow, or grove appeals to each man" (16). Among the Germans, fraternal and paternal ties appear to have been weak. In some cases, for example, fathers even had the right to disavow or sell their own sons. Tacitus even suggests the existence of a matriarchal society ("Sisters' children mean as much to their uncle as to their father: some tribes regard this blood-tie as even closer and more sacred than between son and father...") (20). For more on the Germanic family, see Cuvillier (1986).
[45] At least in the Iberian peninsula, repeated Berber invasions during the Middle Ages ended up emphasizing this presence.
[46] The extremely early female age at marriage in southern Europe (17-19 years of age) during the fifteenth and sixteenth centuries, together with a fairly low incidence of remarriage among women, would seem to have much in common with marriage patterns in northern Africa. Even much later during the eighteenth century, female age at marriage continued to be noticeably earlier in much of southern Europe than in the northern part of the continent. It is worth noting here that in the southern parts of Spain, Italy and Portugal age at marriage throughout the pre-industrial period was always much younger than in northern parts of those same countries. This suggests that even in southern Europe the degree of implantation of these family forms was quite heterogeneous. For Italy, see Da Molin (1990); Delille (1985); for Spain and Portugal, see Rowland (1988).
[47] A number of these ideas towards marriage, especially in that it was not a sacrament and that it rested on mutual acceptance, were originally developed by the Cathars in twelfth-century France (Goody, 1983). For the influence of the Reformation on marriage and attitudes towards family life, see Goody (1983).
[48] On this subject, see Bainton (1952). Regarding the control of sexuality during this period, Goody maintains that 'Eastern' or 'Eurasian' patterns prevailed in southern

Europe. These were based on pressures from a more extensive and more effective kin network as well as on an earlier age at marriage for women, all of which meant that such control was more intensely felt and more easily maintained (1983). These controls were never as effective in Protestant Europe where marriage was later and there was no southern heritage of strong families. On this subject, see also Lebrun (1986).

[49] A good example of these differences can be found in the gap between literacy rates in northern and southern European countries. Even though both sexes showed higher literacy levels in the North, the north/south differences were far greater for women. As late as 1887 in Spain, 70 percent of adult women continued to be illiterate, and in certain areas of the country this percentage was closer to 90 percent (Reher, et. al., 1993). Spanish and Italian women did not reach German female literacy levels of 1700 until after the turn of the twentieth-century, and Swedish levels of 1700 were not reached until the 1960s (Núñez, 1997).

[50] While tackling this issue from the same perspective, Jack Goody (1996a; 1996b) has always tended to maintain a certain degree of skepticism as to the real advantages of the Northwest European family system for processes of economic growth and modernization. He feels that perhaps they were more significant "...in the shape that economic development took rather than development per se." (1996a).

[51] Here, the exaltation of Luther of common occupations as a 'calling' is essential because an individual's job became also his religious calling. In the words of Roland Bainton: "The term *vocation* was transferred by Luther from the cloister to the workshop." On this subject, see the classic essay of Max Weber (1930) *The Protestant Ethic and the Spirit of Capitalism*. See also Bainton (1952).

[52] For comparative East-West viewpoints, especially insofar as they affect subsequent economic development, rational thought processes and family systems, see Goody (1996a, 1996b).

[53] In 1990/91 lone-parent families represented 8.6 percent of all families with children under 18 years of age in Spain, as opposed to 15.7 percent in Germany, 16.8 percent in Canada, 22.0 percent in Denmark, 11.9 percent in France, 18.1 percent in the Netherlands, 22.3 percent in Sweden and 23.5 percent in the United States (Cordón and Tobío, 1996). See also Hantrais and Letablier (1996).

[54] See Cordón and Tobío (1996) and Tobío and Fernández Cordón (1996).

[55] When only never-married mothers are included, the percentage living with the grandmother rises to 60 and 69 percent respectively. It is instructive to note that in northern Spain, especially Catalonia and the Basque Country, the 'grandmother effect' appears to have been much weaker than in the other more southern regions of the country.

[56] According to a 1993 survey, over 35 percent of all persons above 65 intervene on a regular basis in helping their children and grandchildren. This help is often, but not always, centered on child care. See Tobío-Soler and Fernández-Cordón (1996).

[57] An exception on this point is the role of grandmothers in African-American families in the United States where there often represent the survival of the traditional African extended family within a context of the breakdown of the conjugal family.

On this subject, see for example, Wilkinson (1994), Timberlake and Chipingu (1992).

[58] The tradition of civil association is much stronger where government intervention is weak and relatively distant, like in the United States, than where it is ever-present, as in Nordic countries.

[59] The responsibility of the State as opposed to that of the family in supporting the elderly is a frequently debated issue. For historic contexts see, for example, Kertzer (1995).

[60] For a perspective on this issue based on a microsimulation study of kinship networks in Spain, see Reher (1997).

[61] In a recent paper, Anton Kuijsten (1996) has made a strong case that family patterns in Europe is more a case of divergence than one of convergence. For a review of the common demographic constraints affecting family life in Europe, see Bégeot and Fernández-Cordón (1997).

GIUSEPPE A. MICHELI

KINSHIP, FAMILY AND SOCIAL NETWORK

The anthropological embedment of fertility change in southern Europe

1. A METHODOLOGICAL PREMISE

One of the final waves of cholera to hit London in 1854 led John Snow to take the plunge and find out why. His explanation lays on the different qualities of water provided by different companies in various parts of London (Lilienfeld and Lilienfeld, 1980). Snow's experimental plan to discover which water company carried the infection is part of the history of epidemiology as an induction-based science. However, there was no basically coherent etiological model corresponding to this tenaciously followed intuition that water was the place to look for the cause of the disease and not the equally considered alternative place, the miasmatic air. Snow inferred the existence of a 'cholera poison' transmitted to the population via the water from the mouth of the Thames. Another quarter of a century was to pass before microscope techniques developed sufficiently to permit scientists to isolate the 'cholera vibrio' and thus to work out the cause of the contagion. All the same, Snow's use of 'romantic' epidemiological interpretative categories does not detract from the importance of his insight. It is actually because of this that epidemiologists started reflecting on the channels of contagion – even without a clear or systematic theoretical basis.

The year after the 1854 cholera epidemic, Frederic Le Play published in Paris the first edition of "Les ouvriers européens" (Le Play, 1855). From then on, till the 1871 'summa' (Le Play, 1871), Frederic Le Play started systematically mapping European regions, using a typology of the organisational models of the household based on two modern variants of the patriarchal ideal type. In the stem-family, continuity is ensured by blood-ties, with one child being singled out as heir general to the home.[1] The unstable (or nuclear) family arises from the union of two autonomous people, and survives just as long as they survive, exerting over the children both a shorter period of care and looser control.[2] Le Play's analysis was much esteemed during his lifetime but quickly lost credence after his death. Emile Durkheim

G. Dalla Zuanna and G.A. Micheli (eds.), Strong Family and Low Fertility: A Paradox?, 77-104.

was soon to start a course of lectures, criticising him on two grounds: "firstly that it is impossible to generalise from the case studies which tell us much about the individual family, but little about the society in which it is placed; and secondly that this 'sociographie microscopique' involved the collection of a mass of uninteresting detail" (Brooke, 1998).

Current population studies have their own puzzles, too. The stagnation of fertility in southern Europe is undoubtedly one of them. Nevertheless, just as John Snow faced his epidemiological puzzle by analysing hydrological data and pointing to water pollution, we could explore similar empirical evidence: there is considerable overlap between Le Play's mid-eighteenth-century household model map and the regional TFR map of central-southern Europe in the 1980s. The under-valuation of Le Play's work comes from scorning a non theory-laden 'sociographie microscopique' which, furthermore, is used as propaganda for a Vendéean philosophy of life.[3] Yet, like Snow, although unable to give an acceptable explanation, Le Play probably hit upon a fundamental disparity in social and demographic behaviour in Europe.

Of course, in the absence of an interpretative model, the persistent disparities in Le Play could be dismissed as mere statistical coincidences. In this chapter I want to tread another path in two different stages. In paragraphs 3–6 I propose to examine closely the overall structure of relationships involved in Le Play's typology, going beyond the household category and trying to include the networks of both kinship and extra-kinship strong ties. This will lead to formulating a hypothesis of a tri-partite model for western European relationship models. The concluding paragraphs 7 and 8 provide some rough contributions to an etiological model in which the current diversity in regional fertility behaviour is explained by basic persistent anthropological structures. But in order to understand this logical connection we need further premises.

2. DEMOGRAPHIC PRACTICES ARE SPATIALLY EMBEDDED *LORE*

We feel it essential to formulate a more comprehensive theoretical framework of recent fertility changes in Europe; but why on earth is it necessary to expand our analysis beyond the circle of household relationships to include the larger circles of both kinship and non-kinship strong ties? We can justify this argumentation by reflecting that in recent decades the scenario of uniform evolution of demographic patterns, gradually spreading from north to south throughout Europe, seems more and more to conflict with the evidence of a bipolar Europe.

No doubt both in northern Europe and the Mediterranean countries present demographic transformations are the result of the same general process of modernisation. Lesthaeghe (1991) defined a second demographic transition as "a further, much more public, manifestation of individual autonomy (...), more pervasive as it is directed against all expressions of external institutional authority". The family is a major agency of social reproduction, and it is being affected and undermined by this wave of modernisation. All the same, the charge against the institutional authority of the family has acquired different forms in different situations.

If we examine the total fertility rates in continental Europe for the years 1983 and 1993 at the regional level, we realise that Europe is roughly split up in three different areas by two boundary lines running along the 42nd and 47th parallels of latitude north (Micheli, 1996).[4] While northern Europe shows a renewal of fertility rates and the Mediterranean countries (Spain, Greece, southern Italy) a sharp fall, a critical belt between the two parallels (with TFR steadily below 1.5) includes northern Spain and Italy, some Pyrenean and Mediterranean French regions, some German *Länder*, and looks as if it infringes on Slovenia (however not measured by Eurostat data) through the Austrian and Friuli corridor. Cleavages in European demographic behaviour do not respect national boundaries, but rather pass through and into the countries.

Analysis of the total fertility rate on the regional level splits the map of Europe into three rather than two developmental patterns. Even though the plot thickens, the theoretical issue remains unchanged, and it would be easier to begin by facing it in its dichotomous version: if a single macro-process of modernisation is profoundly transforming western societies, whatever their development path, why do the changing mechanisms of intergenerational relations cause a pattern of family break-up in the north and a drying-up of the family in the south? How can the same agency produce quite different demographic patterns?

In order to contextualize these historical variants, it might be useful to rediscover a neglected sociological rule of Durkheim's (1895): if several equally determining (equi-final) processes produce the same result, really the results are similar but not identical, as they have behind them different epigeneses.[5] Consider the example given by Durkheim himself: "In the common sense view, fever designates a single pathological entity; however science classifies more specifically different fevers, with respect to different effects". On the basis of these arguments Durkheim confutes Mill's and Weber's equi-finalistic rule (which leads to "vaguely assigning a badly defined consequence to a hazy and undefined group of antecedents") and formulates the following statement: "A single cause always corresponds to the same effect. If, for instance, a suicide is determined by a number of causes, that happens because we find ourselves faced with different kinds of suicide" (*ibidem*).

Let us cross-tabulate the country-level proportion of extramarital births (as a proxy for the spread of the marriage bond) with the total fertility rate (as a proxy for the spread of a full motherhood experience). It is a well-known fact that, behind a common process of convergence to a standard pattern of demographic rates, European countries follow two distinct demographic 'development paths', hinging upon two distinct mainstays (table 1): the marriage contract without children and numerous offspring without marriage.

Table 1. Sixteen European countries by 1990 proportion of extramarital births and TFRs.[6]

	% extramarital births		
TFR	< 10%	10% - 30%	30% - 50%
1.25-1.50	Greece, Italy, Spain	Austria, (West) Germany	
1.50-1.75	Belgium, Switzerland	Netherlands, Portugal, Scotland	Denmark
Over 1.75		England, Finland, France	Norway, Sweden

Paraphrasing Durkheim we can say therefore that, if the path called a 'second demographic transition' is affected by more than one intervening process, that means there are a number of different 'second demographic transitions'. Actually, with a few broad strokes we can trace a main boundary line in Europe. In northern Europe, demographic transformations took the form of 'charge against institutionalised marriage', i.e., against the horizontal one of the two bonds that the family hinges upon. By contrast southern Europe seems to be characterised by the crisis and break-up of the intergenerational kinship agreements and of the vertical parenthood bond. Motherhood loses its appeal not as the experience of only one child (easily compatible with a full working career) but as an irreversible life choice. Two different and in many ways opposite processes (saving the marital bond at the expense of the ancestral and vice versa) have produced the same result for decades: a decline in Europe's fertility. This has led the researchers to a uniform reading of the processes, throwing them off guard when the trends started to bifurcate.[7]

At this point we must ask another question: what justifies the development in Europe of different epigenetic processes leading to fertility decline? Both current theoretical frameworks (focusing the former social and economic conditions, the latter cultural models) are one-sided and incomplete. Only by connecting one with the other can we find a less partial explanation: the linking thread might be the set of relationships translating social action into social practice and norm. My aim is to reconstruct – both by analytical arguments and by reference to various sources of empirical data – the framework of practices stratified in time, which make up the anthropological embedment[8] of current fertility dynamics.

Table 2. First and second quartile of regional TFR and area of Le Play's stem-family. Spain, 18 Comunidades autonomas; IN = index numbers within the country.

	Le Play area	TFR 1983	TFR 1988	TFR 1993	1983-93 Average	IN	Quartile
Spain		*1.794*	*1.430*	*1.246*	*1.490*		
Asturias	Yes	1.474	1.018	0.848	1.113	74.7	I
Pais Vasco	Yes	1.456	1.101	0.958	1.172	78.6	I
Castilla-Leon	Yes	1.622	1.163	1.000	1.261	84.7	I
Aragon	Yes	1.532	1.231	1.080	1.281	86.0	I
Galicia	Yes	1.634	1.213	1.049	1.299	87.2	I/II
Catalonia	Yes	1.390	1.335	1.226	1.317	88.4	II
Navarra	Yes	1.578	1.278	1.144	1.333	89.5	II
Cantabria	Yes	1.805	1.231	0.998	1.345	90.3	II
Rioja	Yes	1.806	1.234	1.065	1.369	91.9	II

Practices (and norms too) refer to one or more reference actors or groups, and generally (even in the era of globalisation) groups tend to be rooted in a territorial niche and in a subculture or 'folk-lore'. Groups – Carl Schmitt (1963) would say – are 'telluric' actors. Of course, many processes may concur in this geographical rooting, stratified along the latitude, but I am interested in studying a particular sort of social feedback we observe today: while historically different social practices gradually crystallised in the shape of different inertial anthropological structures (norms and values), these in turn embed the current social transformations (whatever economic, political or technical factors cause them) into different new patterns of social practices.

How can we identify these "folkways and customs" (Sumner, 1906) that act as incubators for divergent paths of development? In my opinion, we will never understand the dynamics of the family if we confine ourselves to monitoring only the restricted household circle without exploring the fundamental interplay between the household and two other circles round it: kinship and the network of friends, neighbours and all other strong ties. My hypothesis is that the overall regional patterns of these three circles could influence local differences in social and demographic reproduction strategies.

3. HOUSEHOLD PATTERNS IN HISTORIC EUROPE AND THE CURRENT DEMOGRAPHIC CHOICES

As I said, Le Play does not confine himself to an abstract typology of household patterns: he locates them minutely on the regional map of Europe. In the geography of Le Play (recently recovered and systematised by Todd, 1983) the stem-family area includes the northern and Pyrenean regions of Spain,[9] Pyrenean[10] and Mediterranean regions of France (Languedoc and Provence) and central-northern regions of Italy.[11] It is surprising to note how closely, in the three most populous countries of south-continental Europe (France, Spain and Italy), Le Play's stem-family map and the map of current fertility stagnation overlap. Let us classify these regions (Eurostat data, NUTS level 2) in compliance with the rank order of the total fertility rates 1983-1993 within each country: in Tables 2-4 we found that all Le Play's stem-family regions are located above the line of the median national value.

Table 3. First and second quartile of regional TFR and area of Le Play's stem-family. France, 22 Régions; IN = index numbers within the country.

	Le Play area	TFR 1983	TFR 1988	TFR 1993	TFR 1983-93	IN	Quartile
France		*1.785*	*1.806*	*1.655*	*1.749*		
Limousin	Yes	1.495	1.469	1.347	1.437	82.2	I
Midi-Pyrénées	Yes	1.529	1.552	1.470	1.517	86.8	I
Auvergne	Yes	1.631	1.565	1.389	1.528	87.4	I
Aquitaine	Yes	1.588	1.595	1.445	1.543	88.2	I
Poitou-Char.	Yes	1.704	1.655	1.488	1.616	92.4	I
Languedoc	Yes	1.680	1.699	1.579	1.653	94.5	I/II
Burgundy	No	1.764	1.756	1.582	1.700	97.2	II
Centre	No	1.754	1.775	1.593	1.707	97.6	II
Alsace	No	1.699	1.759	1.651	1.703	97.4	II
Corsica	No	1.872	1.716	1.526	1.705	97.5	II
Provence-Alpes	Yes	1.716	1.790	1.639	1.715	98.1	II

Extending the analysis to neighbouring countries – where Le Play's exploration could be less analytical - does not radically change the framework. This is particularly true for the Swiss *Cantons* and for Austria, where the southern-eastern regions,[12] together with Friuli in Italy, form a bridge to Slovenia, as in Le Play. The result is a sort of southern European orographic ridge: it unfolds from west to east along the cordillera, the Pyrenees, the Cevennes, the Alps and the Apennines. It is the breeding ground both of stem-family culture and drastic current changes in reproductive behaviour.

Table 4. First and second quartile of regional TFR and area of Le Play's stem-family. Italy, 20 regioni; IN = index numbers within the country.

	Le Play area	TFR 1983	TFR 1988	TFR 1993	1983-93 Average	IN	Quartile
Italy		*1.508*	*1.362*	*1.255*	*1.375*		
Liguria	Yes	1.055	1.006	0.972	1.011	73.5	I
Emilia-Rom.	Yes	1.078	0.976	0.979	1.011	73.5	I
Friuli-V.G.	Yes	1.121	1.047	0.968	1.046	76.0	I
Tuscany	Yes	1.186	1.106	1.029	1.107	80.5	I
Piedmont	Yes	1.201	1.102	1.058	1.120	81.5	I
Aosta Valley	Yes	1.303	1.232	1.012	1.182	86.0	II
Lombardy	No	1.288	1.161	1.111	1.118	86.3	II
Veneto	No	1.307	1.184	1.111	1.201	87.3	II
Marche	Yes	1.386	1.231	1.145	1.254	91.2	II
Umbria	Yes	1.441	1.255	1.316	1.276	92.8	II

It is more difficult to use the stem-family area map, which La Play drew up for Germany. According to Le Play,[13] the southern European orographic ridge actually extends over the whole of the Danube basin (Baden-Württemberg and Bavaria) and from there it follows a south-north line running through the Rhineland and reaching Denmark via Hannover, Lüneburg and Schleswig-Holstein. The stem-family area therefore seems to cover a large part of western Germany: other factors can explain internal distinctions.[14] As a proof, the change in the capitalist spirit, described as a reduction in the time horizon of the family home,[15] is found by Schumpeter (1943) at the core of the region where social and political scientists nowadays sometimes locate the German (or Rhenish) "variants on the conservative welfare model". However, this east-west demarcation line does not significantly find a match in the rank-ordering of the 40 *Regierungsbezirke* according to the TFR. This could be because its effects are largely swamped by the consequences of the GDR's political breakdown. The collapse of the TFRs of the *Regierungsbezirke* of eastern Germany could be hiding the previous tradition of lower fertility in the Rhine and Danube areas, compared with the Elbe region.

4. THE CO-ORDINATES OF LE PLAY'S HOUSEHOLD TYPOLOGY

Curiously, since the hypothesis of more distinct household patterns in Europe again attracted the attention of social scientists, Le Play's contribution was rediscovered but also underestimated or misunderstood.[16] In 1990 Hollinger and Haller (1990), confuting the hypothesis of the nuclear family type as the dominant type in all advanced industrialised countries, said: "modern historical family research has disproved convincingly the earlier assumption of the predominance of the extended 'stem' family (Le Play)[17] in pre-industrial Europe". But such an assumption is hardly attributable to Le Play. In fact the authors go on describing analytically three 'European cultural areas' with different family structures in pre-industrial times: and two of these three areas are similar to the Le Play's unstable and stem family.

At least two arguments explain the recurrent misunderstanding of Le Play. The first one, as already stated, is the disaggregation level the authors use for their territorial analysis: passing from country to regional level is enough to trace clearly the cleavage between the area of the stem-family and fertility stagnation and the world of unstable families and fertility recovery (see Reher in this volume). A second argument concerns the lack of clarity in the definition of the stem-family and more generally of Le Play's family typology.

It is common opinion (Todd, 1983) that the stem-family category is founded upon two co-ordinates: a) the degree of neolocalism (liberal model) or patrilocalism (authoritarian model) in the residence at marriage, and b) the rules of inheritance. Nevertheless Caroline Brettell (1991) has confuted the prevalent hypothesis among historians and anthropologists, arguing (at least with respect to the Italian case) that "inheritance practices are not determinative" in discriminating the family models. Behind a lifestyle, Brettell suggests, we should glimpse both an economic calculus and a 'philosophy of life'[18]. Behind a social practice producing social norms, we can see either a system of costs and benefits or a system of values (a meaning-giving system) which has sedimented in time. As Reher says, "historically the strength of familial ties appears to have conditioned the way in which succession was carried out in stem-family regions".

Table 5. Typology of family patterns in Europe (Laslett, 1983)

% complex families	Neo-local residence	Area	Countries
Nearly zero	Predominant pattern	Western Europe	England, Netherlands, northern France
Low	Widespread pattern	Central-western Europe	Southern France, partially Germany
Medium	Minority pattern	Mediterranean Europe	Spain, Italy, partially Portugal, Balkans
High	Almost unknown	Eastern Europe	Russia

In Le Play's typology, then, the rule of inheritance is perhaps subordinate to the rule of the placement of the residence after marriage. As a consequence, the basic cleavage among family models divides patrilocal (patriarchal or stem) and neo-local (unstable) families.[19] Cross-tabulating this variable with the frequency of complex (extended and multiple) households, Laslett (1983) proposed a typology of forms of organisation of the home in traditional Europe (Table 5) that overlaps Le Play's map, and where the crucial cleavage distinguishes a north-western area, characterised by predominant neo-local residence, from a large and compound stem-family area, where the neo-local residence is less widespread or a minority, even though present, practice.[20]

Moving from a country-level analysis to a regional (NUTS 2) one, Laslett's typology can also become inaccurate and the demarcation lines already mentioned in §3 appear again. For instance, Rowland (1988) shows how placing the whole of Spain in the Mediterranean area could conceal the peculiar stem-family culture of the Cantabric-Pyrenean area. As for Italy, Barbagli (1991) has constructed a more analytical typology (Table 6) where southern Italy is – coherently with Le Play – placed into the unstable and neo-local family area, whereas a further cleavage between north-western and north-eastern Italy is not found in Le Play.

To sum up, we are a long way from understanding clearly what the stem-family really marks, but it is geographically unquestionable that the stem-family has its own specific identity. The satisfactory overlap between the maps of traditional household patterns and of current fertility decline confirms the nexus between current changes in demographic behaviour and the persistence of some anthropological structures and practices concerning the formation of the family. A similar result is found by Holdsworth (1998), who traced the Spanish regional map for the age of transition to adulthood.

Table 6. Typology of family patterns in Italy (Barbagli, 1991)

Residence at marriage	Age at marriage		
	Early	Delayed	
Neo-local	Southern Italy	Cities and towns of central and northern Italy and Sardinia	
Patrilocal		Rural central and northern Italy	
		Western Italy (stem-family, small-scale peasant land ownership)	Central and east northern Italy (multiple households – horizontal as well as vertical)

But the influence of anthropological embedment goes beyond strictly demographic behaviour. The modern relevance of the Le Play cleavages (either within the European map or simply within the boundaries of only one country, like Italy) can be extended to other facets of social reproduction. E.g. the compound geography of the regional neo-local and patrilocal family model, described in Barbagli's typology, is perfectly reflected in the map of the architectural forms of farmhouses in Italy.[21]

Moreover, the central-northern regions of Italy marked by the traditional predominance of the stem-family also experienced in the 1980s the upsurge of a new kind of capitalism, the Marshall 'industrial districts' (Piore and Sabel, 1984), hinging upon a network of 'family- firms' managed by a group of sibs, exactly as in Le Play's sketch of the stem family: "(the other children leaving the household) can in turn both become totally independent of each other or embark together on some enterprises..". Italian studies into the 'informal economy' have underlined some crucial qualities of the family-firm: its ability to cope, using non-standard strategies, with all the tensions emerging in a changing society, its autonomy of organisation that makes it a perfect mechanism of crisis management, its resources of flexibility. All this extra-skill of functional adaptation, not only to changes in the social system but also to transformations of the productive system, seems to be a distinctive feature of the stem-family, or – as anthropologists (Linton, 1936) have already noticed – of the consanguineous family.

Finally, it would be useful to reflect on the fact that the orographic backbone of the stem-family (from the Basque country and Catalonia to the central-European areas of Bavaria, Carinthia and Slovenia) contains the core of Europe's family-based, highly ethnocentric 'little homelands'.[22] These are the very regions in which 20th century history sometimes has seen civil wars, i.e., break-up of a social order based on blood ties (Micheli, 1999a).

5. KINSHIP & STRONG TIES: CONCENTRIC CIRCLES ROUND THE FAMILY

All these connections between the Le Play cleavages and other social, economical and political processes make one suspect that household cannot be taken as the only framework for current social changes. Let us try and enlarge the analysis to other dimensions of social relationships. Forty years ago Elizabeth Bott (1957) underlined a chief difference between 'small-scale (primitive) societies' and 'urban industrialised societies':

> "In England and other Western European industrialised societies work groups are seldom recruited on the basis of kinship, individuals may earn a living without depending on relatives for their means of livelihood, productive resources may be owned by individuals who are not related to one another (..). This reduced importance of kinship in economic affairs is associated with a narrower range of kin recognition, with absence of corporate groups of kin (..), with less frequent and intense contact among relatives".

No doubt the model featuring a large range of kin recognition, the importance of kinship and corporate groups of kin in economic affairs is dominant in many non-western societies.[23] But it is also very close to the Mediterranean model of the family-firm and stem-family. Elizabeth Bott's intensive study (Bott, 1957) into a small number of London families and their social networks clears our mind of the "commonplace of sociology and anthropology, that kinship does not play a very important part in industrialised societies". Litwak and Szelenyi (1969) stressed later that the growing thinning out, in western societies, of face-to-face contacts does not mean a loss of importance of the primary groups.

The co-resident household must therefore be analysed as the core of an integrated system, surrounded by at least two circles that are analytically distinct. The first one includes that part of the kinship that is operationally or symbolically close to the household. The second circle, which develops round household and kinship and can extend beyond them, consists of the network of subjects connected with members of the household by strong ties, i.e., 'frequent ties, giving emotional or instrumental support'.[24] If some processes of anthropological embedding, concerning family models, are connected with the current differentiation of demographic behaviour, something similar is also true of kinship and network patterns.[25] Some sort of systemic connection links together the forms and sizes of the three relationship circles.

The first seminal rule of linkage between the family pattern and the connectedness of family networks[26] was formulated by Bott (1957), who distinguished two kinds of families (segregated conjugal[27] versus joint conjugal role relationship[28]) and two kinds of social networks around a family (a 'close-knit' network, with many relationships among the component units, versus a 'loose-knit' one, with few such relationships). Bott concluded: "the degree of segregation in the role-relationship of husband and wife varies directly with the connectedness of the family's social network". In other words, the asymmetrical family appears to have a more dense strong-ties network, i.e., a social network where there are more kinship ties than strong ties with non-relatives (to the extent that the proportion of kin in the network can be taken as a good proxy for its density):

> "Kin are of special importance in any type of network. First, kin are especially likely to know one another, so that the kinship region of the network is likely to be more close-knit than other sectors. Second, relationships with and among close kin are relatively permanent" (Bott, 1971).

An intriguing result emerges from European sociological research: as for both Le Play's household patterns and Bott's 'family and kinship' patterns, the size and form of the strong-ties networks are not homogeneous throughout Europe. Hollinger and Haller (1990), emphasising from the 1986 survey of the International Social Survey Program[29] significant differences between the north-western and central culture areas of Europe and the southern ones, show that the closer the family structure and higher than elsewhere the frequency of face-to-face contacts with kin,[30] the fewer are the strong ties with non-relatives. To sum up, "the importance of friends in people's social support networks is inversely proportional to the importance of extended kin".

Truly, keeping the current habit of identifying Bott's asymmetrical family with Le Play's stem-family, in the contribution of Hollinger and Haller, there is evidence of a sort of paradox:[31] contrary to the inverse relationship between social support networks and the importance of extended kin, which was found in the other six countries surveyed, Italy (the only Mediterranean country in the International Social Survey) is characterised both by a strong presence of kin and a large circle of non relatives strong ties. Nevertheless, if the Italian family is evolving toward a nuclear form but retains marked role asymmetry (Palomba and Sabbadini, 1993), we should expect (in keeping with Bott's and Hollinger and Haller's more general rule) a smaller network of non-kinship strong ties.

Recently some social surveys in the Netherlands (Gierveld, Tilburg and Lecchini, 1995) and in a few regions of central-northern Italy[32] have made it possible to compare the size and form of the network of 'emotionally and/or instrumentally significant' ties among older people (over 65 years old) in the two countries. The networks came out as very different, both in size (14.5 members in the Dutch network, only 5 in the Italian one) and composition.[33] But the small size of the strong tie circle is not peculiar only to older people.

In a recent Italian survey, called the 'Social Barometer',[34] the age-specific curve of network size (Micheli and Billari, 1998) has a parabolic shape, with the smallest sizes at the extreme ages, both among young people (where the network belongs mostly to the outer, non-kinship circle) and among older ones (where the strong ties begin concentrating into kinship). However the peak of network size in full adulthood remains clearly inferior to the average Dutch levels. And such a restricted network is also an intrinsically kinship network.[35] All this is fully coherent with Bott's typology, confuting the old stereotype of an Italian family surrounded by a circle of countless friends and relatives.

The general rule is then confirmed again, even though its rationale escapes us. Why is Europe split up into two patterns of social interaction, with different proportions and roles among co-resident relatives, non co-resident kinship and non-kinship ties? Socio-cultural interpretations stress the role played by the cleavage between private-oriented and public-oriented societies,[36] and the statistical linkage between public orientation and basic education spread can support the 'second demographic transition' hypothesis, i.e., an underlying process of 'modernisation' from north to south of Europe. Actually, the clear preference expressed by young people for non kinship strong ties and therefore the over weighting of the outer circle in their network, as results from Social Barometer data, could also be an effect of the spread of high-school education.

Nevertheless, the difference between network size and density in Italy and the Netherlands is so great that it is difficult to support a simple hypothesis of uniform change in Europe. Moreover the diffusionist approach, unless it unrealistically assumes the absence of a southern European family model, simply refers to some previous process of historical formation of the cleavage.[37] And however it is interpreted, it cannot itself be based upon family models alone.

6. FAMILISM AND ASABIYYAH. TOWARDS A TRIPARTITE TYPOLOGY OF FAMILY MODELS

Can we assign a specific identity to the Mediterranean social interaction model, which is located by Le Play in the unstable or neo-localist family model (close to northern-European countries), but unlike them is marked by asymmetrical roles in the partnership and dense and kinship dominant social network?

The geography of family structures and practices sometimes intertwines with the geography of social practices and cultures, producing new problems. The economic debate about industrial districts has opened yet another classic issue of political science: the civic culture too (the ethos of the community good taking priority over family affairs) has its own geography. Apart from Tocqueville's America, the civic culture is predominant in northern-western Europe, but Robert Putnam (1993) found it in Italy also, just in the stem-family regions.

Familism is the opposite of civic culture: the ethos where the good of the family takes priority over community needs. Familism also has its own geography; Banfield (1958) described it as the prevailing ethos in southern Italy. Dalla Zuanna (in this volume) summarised the nature of familism in the following three points: 1) most people arrange their own life based on the family, both as household and as kinship; 2) the individual utility function is overwhelmed by the utility function of one's own family; 3) society is organised in such a way that strategies based on individual utility are less successful than strategies based on family utility. The first point clears up the link between the familist culture and the underlying anthropological structures of stem-family and close-knit network. The two latter points, tracing the functioning strategy of familistic home, clarify a possible link between anthropological embedding and the current bent for strong demographic restraint:

> "In a familistic society engaged (..) in processes of upward social mobility, an (additional) child is a very heavy burden. In a long period horizon (..) familistic parents want their children to have a social status higher than or at least equal to their own status. Therefore the familial investment is very strong (..) To sum up, a well-off familistic society generates few children because it invests too much in those children" .

Dalla Zuanna's interpretation of fertility decline in Italy is intriguing, because it supports the hypothesis of familism as a normative framework. However, it appears to depend upon the hypothesis that familistic culture is uniformly spread throughout Italy: "the recent fertility decline can be interpreted if we assume familism as the background noise of the whole of Italian society..". How can we reconcile this hypothesis of overall uniformity in family strategies both with Le Play's different family models and with the different family cultures generally attributed to the south and to the centre-north of Italy?

Table 7. Strong ties by age classes in Italy: total number.

Geographical region	15-24	25-34	35-44	45-54	55-64	65-74	> 74
North-west	6.44	6.34	6.83	6.36	5.64	5.06	4.95
North-east	5.63	6.99	6.89	6.59	6.14	6.21	5.00
North-centre	5.94	6.28	6.92	6.49	6.30	5.68	4.89
Centre and Sardinia	6.15	6.87	6.79	6.31	6.35	6.29	6.00
South and Sicily	5.76	6.25	6.92	6.53	6.07	5.64	5.07

Table 8. Strong ties by age classes in Italy: household members.

Geographical region	15-24	25-34	35-44	45-54	55-64	65-74	> 74
North-west	1.61	1.37	1.64	1.48	1.13	0.81	0.67
North-east	1.49	1.45	1.67	1.65	1.26	1.00	1.00
North-centre	1.65	1.32	1.68	1.55	1.26	0.91	0.95
Centre and Sardinia	1.52	1.44	1.47	1.82	1.42	1.21	1.00
South and Sicily	1.68	1.46	1.58	1.80	1.48	1.09	0.57

Actually, the division of the social network of Italian families into its three main components (household, non co-resident relatives and non kinship strong ties), measured by the Social Barometer, gives some evidence of a slightly different geography than Le Play's maps (Tables 7-10). The frequency of household components among the strong ties is higher in the regions of central-southern Italy, as a consequence of current delayed demographic transition. The frequency of non co-resident kinship in the network (proxy for the close-knit network of Bott's model) is systematically higher, in 40-year-olds and over, in central-northern Italy (core of the stem-family area). On the other hand, the frequency of non-kinship strong ties is high throughout the life course only in north-eastern Italy, i.e., principally in the Veneto, the only probable location of the unstable, loose-knit network family.

So, if we want to understand the rationale behind the geography of family patterns in Italy – and maybe elsewhere – we need two rules that can strengthen our method of analysis. First, we must give up any interpretation of the social and demographic European dynamic at the purely country aggregation-level, but also, perhaps, at the regional (NUTS 2) level. E.g., the two Le Play regimes in southern Italy intertwine and alternate, depending on the local features of urbanisation and productive organisation. Delille (1988) emphasises how the patrilocal stem-family tradition prevails in hilly areas divided into farms and among the urban high classes, whilst the neo-local nuclear family tradition is dominant in the large landed estates and among the urban lower classes.

Table 9. Strong ties by age classes in Italy: non co-resident relatives.

Geographical region	15-24	25-34	35-44	45-54	55-64	65-74	> 74
North-west	2.23	3.15	3.80	3.43	3.32	3.21	3.09
North-east	2.09	3.59	3.78	3.50	3.49	3.58	2.33
North-centre	1.83	3.08	3.62	3.65	3.85	3.87	2.89
Centre and Sardinia	2.13	3.45	3.91	3.26	3.82	3.61	4.50
South and Sicily	1.97	3.13	3.88	3.53	3.41	3.71	3.07

Table 10. Strong ties by age classes in Italy: non-kinship members.

Geographical region	15-24	25-34	35-44	45-54	55-64	65-74	> 74
North-west	2.60	1.83	1.39	1.45	1.18	1.04	1.19
North-east	2.05	1.95	1.44	1.44	1.39	1.64	1.67
North-centre	2.46	1.88	1.62	1.29	1.19	0.89	1.05
Centre and Sardinia	2.50	1.99	1.40	1.24	1.11	1.47	0.50
South and Sicily	2.11	1.67	1.46	1.20	1.18	0.84	1.43

Above all we need to replace the dichotomous typology with a tripartite one to grasp the difference not only between stem family and unstable family, but also between a northern-European unstable family regime and a Mediterranean one. Reher pursues a similar aim: he suggests a process of hybridisation of the stem-family by a different model, which is prevalently based on 'kin ties' and 'extended family loyalties'. If we wish to explore this model, a capital text is at our disposal: in the Mouqaddima,[38] Ibn Khaldun's theory of collective action hinges upon the concept of 'Asabiyyah, i.e., 'esprit de corps', group solidarity. 'Asabiyyah' (Gabrieli, 1930) is the abstract from the noun 'asabah', i.e., male sibs of a common lineage. 'Asabiyyah' is based upon blood bonds, reciprocal aid – the Polanyi reciprocity – produced both by the 'nasab' (genealogy) and indirectly by some non blood ties, such as alliance (hilf) or patronage (wala).[39]

If we attribute to the concept of 'Asabiyyah' the extended meaning of 'alliance among kin',[40] we can easily realise how exactly it looks like the concept of a close-knit network of the Mediterranean area. In describing "small-scale (primitive) societies" Bott (1957) in fact stressed that the elementary (stem-) family

> "Is encapsulated not only within a local group but also, particularly in the sphere of domestic affairs, within a corporate kin group (..). When there are corporate local groups and kin groups, segregation of conjugal roles is likely to become even more marked than that described above for urban families with close-knit networks. Marriage becomes a linking of kin groups rather than preponderantly a union between individuals acting on their own initiative".

Table 11. A tripartite outline of the family models in Western Europe.

	Le Play's neo / patrilocal-ism	Bott's internal role set	Size & density of social net-work	Predominance among strong ties
Atlantic pattern	Unstable family	Symmetrical family	Large, loose-knit network of strong ties	Many neighbours & friends (bridge towards weak ties)
Latin pattern	Stem-family	Asymmetrical family	Small, close knit network of strong ties	Kinship predominance (family-firm)
Mediter-ranean pattern	Unstable family	Asymmetrical family	Small, close Knit network of strong ties	Encapsulated in kinship (families alliance, 'Asabiyyah')

The Italian Social Barometer data show that the ideal type of family encapsulated in its kinship is present both in the centre-north and in the south of Italy. In both these regions the web of affiliations is mostly circumscribed with blood-bond ties, and with few non-kinship ties. Nevertheless, while in central-northern Italy kinship acts as a bridge towards the land of weak ties, in the Mediterranean unstable family the kinship circles of different families are inclined to intertwine with each other.

On the other hand (Table 11) the Mediterranean unstable family and the north-European one are somehow similar, as both are integrated in a larger land of ties. However in the north this is the land of weak ties, whilst in the south – where the extra-kinship network is traditionally shorter (Trumbach,

1978) – a different, but equally effective, network is woven by the intertwining of different kinship systems, which is by an alliance among families. The key of family-centred social reproduction is here, therefore, a policy of kinship: it acts by means of collateral line relationships, which weave different family threads into a single close-knit web of reciprocities.

7. CONCLUSIONS. IN SEARCH OF SYMPTOM-FORMATION FACTORS

Reasoning on the anthropological frame underlying the current demographic transformations in Europe, three main arguments have been developed here. First, the European patterns in fertility decline are regionally embedded 'lore', roughly concentrated in three bands at different latitudes and overlapping Le Play's geography of family models. And this overlap concerns not only demographic dynamics but also both economic and political ones. On the one hand the predisposition of sibs to 'embark on common entrepreneurial initiatives', typical of the stem-family, takes the shape of the 'family-firm' in central-northern Italy industrial districts; on the other the importance of blood-ties in stem-family areas is the incubator of the ethnocentric philosophy of 'little homelands'.

Secondly, the anthropological roots of European demographic cleavages cannot be reduced merely to the household; they are also to be found in the concentric circles of strong ties. Crossing Le Play's family classification, according to external strategy (leaving the family, alliance among sibs), with Bott's family typology, according to segregation in the internal role-set, we find that both the stem-family area and the unstable area in southern Europe are marked by a small, close-knit network of strong ties, with kinship predominance. The Italian Social Barometer data confute the hypothesis of a large Italian network both of relatives and friends, and confirm Bott's findings. Close-knit network and kinship dominance are also structural characteristics of a 'familistic' society, strategically prone to strong fertility control.

Third, the kinship dominance area is not homogeneous within itself: different patterns of social network organisation are to be found in it. In the stem-family area, social support hinges upon a network of kin (consanguineous), whilst in the unstable Mediterranean area social support hinges upon an alliance primarily among different kindred units, then upon a network with many relatives-in-law.

A clear claim for future research emerges from these results: we can no longer avoid investigating the inertial anthropological localisms where today's demographic dynamics are embedded. From this point of view to pose the alternative between converging or diverging European demographic dynamics is only a misleading question. Reher emphasises it, basing his statement upon the valuable category of 'path dependency'; Weber (1904) was working on a similar issue when he wrote: "development paths too can be constructed as ideal types".

But a good sociography is not enough. We must also try to understand the interlocking of structural conditions, rational choice, practices and norms. An open question remains: why, in the present historical circumstances, do we see a drastic drop in fertility behaviour precisely in the regions where the importance of the family agency (in the shape of the stem-family or of familistic kinship alliance) is embedded in the anthropological rules of social reproduction?[41] How can we explain the relationship between family predominance as anthropological embedding and family collapse as demographic reaction?

It may be useful to reconsider these questions in the light of the cognitive dissonance theory. Festinger's theory (Elder and Caspi, 1988) can be summarised by saying that whenever a person holds two or more 'cognitions' (including beliefs and norms, preferences or emotions) that are inconsistent with one another, and the tension produces psychic discomfort and physiological painful arousal, an unconscious pressure is set up to adjust one or more of the elements in the set, by changing or blocking some of the dissonant cognitions, so as to reduce the tension and restore consonance.

There is no doubt that, over the last decade, not only one part but the whole of Europe has come to the zenith of a long period process of homogenisation in the name of modernism and secularisation (Lesthaeghe, 1991). However, we know that this long process has brought to light strong contrasts between economic conditions and expectations about the way of life-.[42] It is equally evident that, in the last two decades of the century, the final outcome of this process is as far off as behavioural standardisation. As we have seen before, northern Europe favours self-fulfilment by procreational choice outside marriage, whereas central and Mediterranean Europe preserves the marital bond at the expense of the maternal blood bond – two opposing strategies which, however, for several years, have both contributed to the same declining trend in fertility.

Faced with such tension between resources and expectations, various parts of the continent have used diverse adaptive strategies. So what 'law-like statement' can help explain why? Analysing the similar problem of changes in philosophy of life among 'the children of the Great Depression',

Elder and Caspi (1988) resort to a social mechanism that they defined as "principle of accentuation".

> "Social change creates a disparity between claims and resources, goals and accomplishments and the corresponding loss of control prompts efforts to regain control. (..) Adaptive responses are shaped by the requirements of the new situation, but they also depend on the social and psychological resources people bring to the newly changed situation. Individual and relational attributes, such as coping styles and the marital bond, affect adaptation to new circumstances. The accentuation principle refers to the increase in emphasis or salience of these already prominent characteristics during social transitions in the life course."

Tesser and Achee (1994) pose the same problem of indeterminacy among different mechanisms to get out of a dissonance situation. They suggest a solution going beyond state variables and introducing "path dependence or hysteresis".[43] According to them,[44] a not very frequent pattern of behaviour, under strong opposite social pressure, tends to further dissipate with time, just as a frequent pattern of behaviour tends to further increase: their thesis is similar to the principle of accentuation.

Also the bifurcation of demographic patterns in Europe may be attributable to the same mechanism. The persistent prestige of the marital bond, in Mediterranean countries, leads not only to protecting it but also to highlighting its importance at the cost of mother-child links. The traditional weakness and instability of conjugal ties in Atlantic countries becomes accentuated by complete dissociation from procreation. The various European regions adapt their own demographic behaviour to mitigate the effects of dissonance by barricading themselves into the fortress of their respective strong cultural specificities.

As a by-product of this thesis, Elder, and Tesser and Achee help us identify some contextual characteristics that can explain the diversification mechanism. What "symptom-formation factors", to use Brown and Harris' terminology (Brown and Harris, 1978),[45] justify the appearance of one 'symptom' rather than another when there is a crisis in a 'body'? Although many processes can concur in these geographical cleavages, stratified according to latitude,[46] both Elder, and Tesser and Achee focus their analysis on the role of practices and norms, i.e., the stratified relational systems, as 'factor-formation' systems.

As a matter of fact we observe today the following sort of social feedback. While historically different social practices gradually crystallised in the shape of different inertial anthropological structures (norms and values), these in turn embed the current social transformations (whatever economic, political or technological factors cause them) into different new patterns of social practice.

8. POST SCRIPTUM. FLOWING BACK INTO THE RIVER-BED OF THE STEM-FAMILY

Applying the accentuation principle, we can assume that different, regionally rooted, family and kinship patterns "react" in contact with an appropriate reagent, such as the macro-process of modernisation, generating different patterns of today's demographic behaviour. In such a way the economic and structural changes in the 1980s and 1990s (with an imbalance between aspirations and resources producing a need for greater control) should have provoked a sort of ebb into the bed of anthropological practices and structures prevailing in the southern and Mediterranean regions.[47]

A reflux, by the way, that very often takes the shape of the effect of a rational choice. We know, for instance, that in twentieth-century cities both the size and form of urban apartments prevented the eldest child from keeping the patrilocal residence at marriage. Nevertheless, the growing well-being of the southern countries of Europe during recent decades has produced a marked increase in the average size of homes. In this more comfortable dwelling system, an only child can again continue living in the family home, even after his marriage, and this is convenient, both for the child and for his parents. So, far from being swept by the northern European family pattern, the stem-family and the kinship-alliance family patterns in southern Europe would seem paradoxically to have been revitalised by contact with the wave of modernisation.

Giuseppe A. Micheli, Institute of Population and Geographical Studies, Catholic University of Sacred Heart, Milan.

NOTES

1. (In the stem-family) "un des enfants, marié près des parents, vit en communauté avec eux et perpétue, avec leur concours, la tradition des ancêtres. Les autres enfants s'établissent au dehors quand ils ne préfèrent pas garder le célibat au foyer paternel.

Ces émigrants peuvent à leur gré rester indépendants l'un de l'autre ou tenter en commun des entreprises, rester fidèles à la tradition ou se placer dans des situations nouvelles créées par leur propre initiative" (Le Play, 1855).
2. (In the unstable/nuclear family) personne ne s'attache à un foyer, les enfants quittent séparément la maison paternelle dès qu'ils peuvent se suffire à eux-mêmes, les parents restent isolés pendant leur vieillesse et meurent dans l'abandon. Le père, qui s'est créé une existence en dehors de la tradition de ses aieux, n'inculque guère sa pratique à ses enfants: il sait d'ailleurs que ses efforts ne sauraient aboutir à un résultat durable. Les jeunes gens s'inspirent surtout de l'esprit d'indépendence. Dans le choix de leur carrière, ils cèdent à leur inclination et aux impulsions fortuites du milieu social qui les entoure" (Le Play, 1855).
3. Emmanuel Todd (Sumner, 1906), who in the 1980s rediscovered the importance of Le Play's contribution to family anthropology, describes him as "aussi heureux dans ses recherches empiriques que pathétique dans ses propositions politiques".
4. Eurostat does not publish TFRs on a regional level (NUTS 2). Therefore we have disaggregated the national TFR, beginning from two sets of available regional data (annual births and distribution of women classed by age), using a method suggested by Gini in 1932 and then recovered by Calot (Brettell, 1991). Comparing the Italian official TFR with the estimates obtained we note the robustness of the method, with a 1% average error (2% for the smaller regions).
5. Durkheim (1895) maintains that Stuart Mill's axiom of a plurality of causes (a consequence does not always flow from the same antecedent; on the contrary it can result now from one cause, now from another one) is the denial itself of the causality principle. No doubt, "if we agree with Mill that cause and effect are absolutely heterogeneous and without any logic link between them", there is no contradiction in the assumption of plurality of causes. Nevertheless if the cause-effect relation acts intensionally and not only extensionally or – Durkheim would say - if it consists of a "natural" relation, "the same effect can have such a relation just with only one cause".
6. Sources: author's calculations based on Eurostat data.
7. Elster (1999), discussing the mechanisms underlying human actions (i.e., "frequently occurring and easily recognizable causal patterns that are triggered under generally unknown conditions or with indeterminate consequences") distinguishes type A mechanisms ("which arise when the indeterminacy concerns which – if any – of several causes will be triggered") and type B mechanisms (which "arise when we can predict the triggering of two causal chains that affect an independent variable in opposite directions, leaving the net effect indeterminate"). The demographic decline from the seventies to the nineties could be classified as specific contamination between both types. The triggering of two, logically self-contradictory, causal chains set off similar effects.
8. Here and below I use the terms embedment and embedding in the sense that Polanyi (1944) attributed to them, to refer to the relation between society and the economy (embedded or not embedded in it).

9. From west to east: Galicia, Asturias, Cantabria and the northern part of Castilla y Leòn, Paìs Vasco, Aragon, Navarra and La Rioja, Catalonia. Holdsworth (1998) circumscribes a similar area (the only significant absence is that of Catalonia.) characterised by a late timing of leaving home for young men, and refers to the Le Play's classification.
10. Midi, Auvergne and Aquitaine, Poitou and Limousin.
11. Except a great part of the plain of the River Po, including Lombardy and Veneto.
12. Salzburg, Tirol, Vorarlberg and Kärnten.
13. In *L'organisation de la famille* (Le Play, 1871) Le Play often lists the regions of continental Europe where the patriarchal and stem-family models prevail. "La famille patriarcale (..) domine sur certaines montagnes (..) notamment sur les hautes prairies des Alpes, du Vivarais, de l'Auvergne, du Jura et des Vosges. Elle se conserve également dans les grandes métairies du plateau central de la France" (§ 7, p.27). "La famille souche offre ce caractère dans les États scandinaves, le Holstein, le Hanovre, la Westphalie, la Bavière méridionale, le Salzbourg, la Carinthie, le Tyrol, les petits cantons suisses, le nord de l'Italie et de l'Espagne. Elle est encore représentée en France (..) dans les Pyrénées françaises et espagnoles" (section 8, p.31). "Les populations slaves et hongroises se groupaient pour la plupart en familles patriarcales sous le régime d'engagements forcés qui a régné parmi elles jusqu'aux réformes commencées en 1848. Elles se rattachent peu à peu à la famille-souche (..). Toutes les races de propriétaires scandinaves offrent dans leur famille-souches d'admirables modèles. En Norvège, en Suède, au Danemark (..). Les familles-souches qui parlent la langue allemande sont mêlées en beaucoup de lieux, près du Rhin surtout, à la famille instable (..). En tête des meilleurs types se placent les paysans du Lunebourg hanovrien (..). Après le Hanovre on peut citer les duchés du nord-est, la Westphalie, le midi du grand-duché de Bade, du Wurtemberg et de la Bavière, la Carynthie, le Salzbourg, le Tyrol, le Vorarlberg et les petits cantons catholiques de la Suisse. Les paysans à famille-souche se conservent avec d'excellentes qualités dans les deux péninsules du Midi. En Italie ils se rencontrent surtout dans le Lucquois, le nord de l'Apennin et les hautes vallées des Alpes. Au Portugal ils résistent encore dans les montagnes du nord-est (..). En Espagne ils luttent (..) dans la Galice, le Léon, les Asturies, la Navarre, l'Aragon et la Catalogne. Enfin dans les provinces basques.." (section 12, p.94 ff.).
14. Federkeil (1997) found "a polarisation between a growing 'non-family' sector on the one hand, which internally is quite heterogeneous or 'pluralized', and a shrinking family sector on the other hand, in which the traditional breadwinner-homemaker is still dominant, although under some attack".
15. "The family and the family home used to be the mainspring of the typically bourgeois kind of profit motive. Economists have not always given due weight to this fact. When we look more closely at their idea of the self-interest of entrepreneurs and capitalists we cannot fail to discover that the results it was supposed to produce are really not at all what one would expect from the rational self-interest of the detached individual or the childless couple who no longer look at the world

through the windows of a family home. Consciously or unconsciously they analysed the behaviour of the man whose views and motives are shaped by such a home and who means to work and to save primarily for wife and children. As soon as these fade out from the moral vision of the businessman, we have a different kind of *Homo Oeconomicus* before us who cares for different things and acts in different ways. For him and from the standpoint of his individualistic utilitarianism, the behaviour of that old type would in fact be completely irrational. He loses the only sort of romance and heroism that is left in the unromantic and unheroic civilisation of capitalism – the heroism of *navigare necesse est, vivere non necesse* (seafaring is necessary, living is not necessary, inscription on an old house in Bremen). And he loses the capitalist ethics that enjoys working for the future irrespective of whether or not one is going to harvest the crop oneself" (Schumpeter, 1943).

16. To tell the truth, Le Play was not the only one, in the mid-1800s, to draft a sociography of the family. There are surprising similarities in Riehl's work: "In 1855, the year in which 'Les ouvriers européens' appeared, the third volume of Wilhelm Heinrich Riehl's 'The Natural History of the German People' was published, a work that considers the particularities of family structures in Germany. Le Play and his German alter ego reach broadly similar conclusions. For both of them, the German family model, the idea type of 'stock family' (Stamm-familie) that could also be found in other Nordic regions and elsewhere in enclaves in Europe, stood in marked contrast to the type of family that predominated for instance in northern France – that is the 'unstable' or nuclear family" (Schultheis, 1999).

17. By a curious *lapsus calami* the authors replace the name of Frédéric Le Play with that of Gustave Le Bon, author in the same years of a "Psychology of the Crowd", another landmark in studies on the mechanisms of social reproduction, which nevertheless has nothing to do with stem-family geography.

18. "One does not have a three-generation stem family because property is transmitted impartibly; one has such a family because parents want at least one child to remain at home, work on the farm, and assist them as they get older. In other words, within the broad context of the law, mechanisms for transferring property are strategies pursued to solve some of the problems faced by families of the past and the present, of Italy, Greece and Portugal – how to secure support in old age, how to contract a marriage for a child, how to provide for all one's children, how to maintain the social status of all members of the family. Transferring wealth is a form of economic behaviour, but as with most economic behaviour patterns studied by anthropologists, it has a social dimension as well. Through the transmission of property people make powerful statements about the meanings of parenthood and childhood, of maleness and femaleness and of kinship and alliance" (Brettell, 1991).

19. Of course, there are other possible readings of the heterogeneity among national household models. In the nineties, some analyses are based on the category of individualistic disposition. E.g., Strather (1992) describes the individuality of people as 'the first fact of English kinship', emphasising that people are treated as unique persons rather than occupants of positions in a kinship universe".Influential is the work of MacFarlane (1978),who traces back the present English model of kinship to a

thirteenth century cultural syndrome he names 'individualism' (independence of children from their parents, kinship ties relatively weak and not linked to a common economy, contractual nature of inheritance). This is a transcultural approach, a sort of collective psychology, which is entirely legitimate but one that I do not use at all in these pages.

20. Le Play's typology, reduced to the single dimension of localism, tends to converge with the criterion suggested by Hajnal (1982), which as a rule contrasts the central European type, compatible with neo-localism, with the non-compatible type.

21. Italian geographers have pointed out that the stem-family area is marked by one-family multi-storey building, the western plain of the river Po by courtyard houses, whilst in southern Italy detached houses prevail, as small one-family one-storey buildings or farmhouses.

22. An example of the close connection between family models and cultural ethnocentrism is the 'pairalist' culture of Catalonia: "There is an ideology of the Catalan family based on 'pairalismo' (the rural house, at once the source of family and tradition) and associated with cultural nationalism. That means that national differences can be expressed in terms of family customs, because the family is related to a particular cultural tradition. In the same way as nation can be expressed as 'casa nostra', the institution of casa is an element of cultural identity and of differentiation with other cultures" (Bestard Camps and Contreras Hernandez, 1999).

23. See for instance the 'pazar' north-African economy, discussed in Boserup (1970).

24. In spite of a twenty-year debate – opened by Granovetter's (1973) suggestions – about the different social weight of strong ties and weak ties, we agree today to identify the latter with acquaintances, but we do not know what should be included in the former. Litwak and Szelenyi (1969) still considered without distinction kin, friends and neighbours as the three main primary-groups. In my opinion a correct taxonomy of strong ties would have to include, besides kinship ties, at least five kinds of ties (not all taken into account by sociological literature): a) ties arising out of the space (neighbours) or time (friendships within a peer group) of every day life; b) alliances of reciprocal solidarity made, in Mediterranean cultures, on the occasion of key life passages (e.g. marriage witnesses and godparents); c) step-relatives acquired by chains of marriages; d) the alliances of reciprocity drawn up among people who have all gone through similar critical life emergencies; e) any other strong tie people can develop in their public life from the universe of weak ties (acquaintances, work colleagues).

25. We cannot go on studying changes and divergences in family models totally detached from changes and divergences in kinship and networks. Over-optimistically Bott (1971) quoted a passage from Harris (1969): "perhaps the really lasting significance of Bott's study is that she has made impossible the proliferation of studies of the internal structure of the family which take no account of its social environment". Unfortunately still today both demographers and sociologists hardly respect this elementary rule.

26. Bott uses the term 'network' in what has come to be called the 'egocentric' sense, "conceptually anchored on a particular individual or conjugal pair", and the term 'connectedness' as synonymous with 'density', to describe the extent to which the people known by a family know and meet one another independently of the focal family.
27. Where complementary and independent types of organisation predominate.
28. Where joint organisation is relatively predominant. Young and Willmott (1957) define as "symmetrical" the family Bott defines as a 'joint conjugal role-relationship'.
29. Including seven countries: Australia, Austria, Britain, West Germany, Hungary, Italy, and the United States.
30. As in Hungary and Italy, where "there is some evidence that socio-cultural factors are out-weighing the influence of modernisation. Even in the highly industrialised northern parts of Italy kin relations are much more similar to the overall Italian pattern of close kin contacts than to the loosened kin contacts of people in north and western Europe" (Hollinger and Haller, 1990).
31. The overlap between Le Play and Bott's categories produces a second problem. Le Play places the south of Italy (and other Mediterranean regions) in the area of nuclear family, so contradicting the evidence of a Mediterranean strongly role-segregated family. The issue is discussed in the next paragraph.
32. In northern Tuscany and in the metropolitan area of Milan (Micheli, 1999a).
33. The Dutch network was larger both in the kinship and in non-kinship components, whilst the Italian network, nearly completely reduced to its kinship components, was further dried up by a below replacement fertility regime going back to the first decades of the century.
34. The Social Barometer was a quarterly survey, carried out by Abacus for two years (1996-1998) over a national sample of about 4,000 interviews, stratified by sex, age, education, size of residence town and geographical regions.
35. Among older people the percentage of non-relatives in the social network is nearly 40% in Netherlands whilst it is only 27% in Tuscany, and it varies from 23% to 30% along the life span in the Social Barometer.
36. "If we assume that private-orientation means social networks consisting mainly of primary group relations and public-orientation means social networks with more secondary relations, (..) Historical family research shows that in the south- and east-European culture area primary-group ties are closer than in north-western-Europe and that the Anglo-Saxon nations have gone even further in the dissolution of kin ties" (Hollinger and Haller, 1990).
37. As for Trumbach (1978), for example, the European family presents two competing forms of kinship organisation as far back as the 11th century. The egalitarian ideology of the 17th and 18th centuries should have spread up over the northern Europe the popular kin recognition system or folkway, where the individual is surrounded by a single network of relatives, including both kin and relatives-in-law, and society is cemented by friendship, patronage and neighbourhood ties rather than by kinship ties.Contrarily, the diffusion over central Europe of some elements of the

aristocratic model of patrilineage or kindred, where kinship is less extended but more central, could explain the rise and placement of the stem-family.

38. Mouqaddima is the methodological introduction to a World History (Kitab el-Ibar) that Ibn Khaldun, historian of the Islamic declining Empire, wrote between 1375 and 1379.

39. "Il existe évidemment dans la nature de l'espèce humaine une disposition qui porte les hommes à s'attacher les uns aux autres et à former un groupe, même lorsqu'ils ne se rattachent pas à la même lignée (..), et la asabiyya qui en est la conséquence engendre seulement une partie des effets auxquels il donne lieu dans ce dernier cas. La plupart des habitants dans une grande ville sont alliée par mariage; ceci entraîne l'intégration des familles les unes dans les autres et l'établissement des liens de parenté entre elles.." (Ibn Khaldoun, 1965).

40. The indeterminacy of the concept of 'Asabiyyah is outlined by Baali and Wardi (1981): "In spite of his great reliance upon the term 'Asabiyyah, Ibn Khaldun never clearly defines it. It seems that the term was quite familiar, or known, in his time; thus he did not feel any need to define it. It may be sufficient for the purpose at hand to define 'Asabiyyah as the tribal loyalty or spirit which make the individual devote himself to his tribe and view the world through its eyes".

41. Le Bras (1999) poses a similar problem: "By a curious paradox in one part of Europe the family is stifling fertility, while in another the importance attached to the mother-child relationship, or its institutional replacement, endorses fertility and so pushes the total fertility rate up".

42. The most notable and most discussed, but not the only one, is that produced by the increased opportunity cost for women bearing children and by the consequent change in women's role.

43. For a formal approach to the diffusion of fertility control, reflecting the "random and path-dependent spread of information in social networks", see Kohler (2001).

44. "Assume that one's disposition is consonant with engaging in the behaviour and that undergoing negative social pressure is dissonant with engaging in the behaviour; one's disposition is dissonant with not engaging in the behaviour, and the presence of negative social pressure is consonant with not engaging in the behaviour. If one starts out high on the behaviour in the face of strong social pressure, then as one's disposition decreases, dissonance increases. To reduce the dissonance, one will look for additional cognitions to support the behaviour. Hence the behaviour will tend to remain high even in the face of a decreasing disposition. On the other hand, starting with strong social pressure and low levels of behaviour, increasing one's disposition will increase dissonance. To reduce the dissonance, one will look for additional cognitions to support not engaging in the behaviour. Hence the behaviour will remain low even though the disposition is increasing" (Tesser and Achee, 1994).

45. Coherently with the Durkheim approach (equifinalistic processes produce similar but not identical results) Brown and Harris (1978), studying the ætiology of women's depression, links vulnerability (background) factors and provoking agents (or events) with a third kind of causal factor, which they name "symptom-formation factors", i.e., factors that "influence only the form and the severity of depression".

An identical symptom (an identical effect) can be developed in a different underlying process (produced by different symptom-formation factors) and then multiplies itself in a range of symptoms with well-distinguished meaning. To understand present demographic processes also requires gathering empirical evidence and sharpening conceptual tools, in order to spot the symptom formation factors, which control the switching over to other possible demographic strategies, as deliberate reactions against situations of cognitive dissonance.

46. The anthropologically embedded practices and norms are the most evident and most widely explored symptom-formation factors for demographic behaviour, though other factors just as promising can be singled out. Exploring several possible outcomes in a situation of cognitive dissonance, Elster (1999) distinguishes autonomous behaviour or mental processes, governed by the reality principle rather than the pleasure principle, and mechanisms that operate at an unconscious level, such as wishful thinking or adaptive preference formation. With regard to the latter, Elster emphasises the absence of a causal model to justify mental strategy going in either direction: "Nothing is known about when dissonance reduction takes the form of wishful thinking and when it appears as adaptive preference formation". Elsewhere (Micheli, 1999a) I have tried to explore the functioning of drives and dispositions, which can form at various crucial phases of the life cycle, as factors motivating or deactivating family choices.

47. Exploring some signals that the youngest cohorts in four countries of central Europe should be exhibiting ideational trend reversal, Lesthaeghe and Moors (1995) concluded: "We are not sure that prospective developments with respect to these issues would be supportive of the 'coming back of the old family'. More likely is that various forms of family formation will continue to coexist, and that the rapid growth period of less conventional family patterns may have come to an end. In short, diversity is likely to prevail in the next decade, but the relative shares of each type may not be changing all that much any more. 'Stability in diversity' seems to be the more appropriate description for the near future".

GIANPIERO DALLA ZUANNA

THE BANQUET OF AEOLUS

A familistic interpretation of Italy's lowest low fertility

1. INTRODUCTION

A number of authors have described how marital and reproductive behaviour in the Western World has changed over the last 30 years (see chapter 1 of Rosina in this book). By way of summarising this tale, the decline in fertility has been accompanied by the ageing of fertility and marriage patterns, increasing cohabitation and extramarital births, along with rising union dissolution and remarriage (or re-cohabitation). In Italy, late fertility and marriage are in line with northern and central European countries (NCEC), whereas cohabitation, divorces and extramarital births are relatively few (see Table 1). Why have some changes diffused easily, whereas other kinds of "modern" behaviour are rare? And what is the cause of the "resultant" lowest low fertility rate?

The low level of cohabitation and extramarital births should be underlined, as in Western Europe, during the 1990s, the more diffused extramarital births are, the higher the TFR is, with a striking change compared to the past, until the beginning of the 1980s (see Figure 1 and Table 2). Empirical data show that in Italy the low cohabitation rate is strictly related to the late departure from the parental home, as the transition from parental home to living alone is rare (Billari *et al.*, 1999b, 2000, 2003). Consequently, to understand Figure 1, the association between the late departure from the parental home and the lowest low fertility rate should be studied.

In this chapter, the familistic characteristics of Italian society are discussed. I will illustrate how familism lubricates the behavioural mechanisms underlying the Italian lowest low fertility. After a description of the Italian familistic way of life, the influence of familism on impeding departure from the parental home and fertility will be described. This chapter deals with Italy, the data on other European countries being used in a comparative way, in order to better emphasise the Italian situation.

G. Dalla Zuanna and G.A. Micheli (eds.), Strong Family and Low Fertility: A Paradox?, 105-125.

Table 1. 12 indicators of marital and reproductive behaviour in 18 western Europe countries. Year 1996 if available or 1995. Rank of Italy in brackets.

	B	DK	G	GR	SP	F	IRE	I	L
NUPTIALITY									
Mean age at first marriage for women	25.6	29.7	26.6	26.0	27.1	27.4	27.9	26.8 (9)	26.5
(Unmarried / women) · 1,000, age 35-44	95.9	195.2	109.1	72.2	95.5	158.4	123.0	106.7 (6)	93.3
Total first marriage rate · 100	56	67	57	52	58	54	59	60 (15)	58
FERTILITY									
(Extram. births/unmarried) · 1,000, 15-44	21.2	53.2	19.2	3.5	9.5	45.5	29.4	7.7 (2)	24.3
(Legitimate births/married) · 1,000, 15-44	80.6	82.4	73.5	78.1	76.1	81.6	96.5	76.3 (5)	96.4
(Births / women) · 1,000, 15-19	6.8	7.5	9.8	9.6	5.9	7.0	13.0	5.1 (3)	7.1
(Births / women) · 1,000, 35-44	16.4	39.5	17.7	16.9	23.4	25.3	44.3	22.3 (6)	28.4
Mean age at first birth	27.1	27.5	28.3	26.8	28.5	27.9	27.4	27.8 (12)	27.5
TFR of women born in 1955	1.83	1.84	1.67	2.03	1.90	2.13	2.67	1.78 (2)	1.69
TFR of the year	1.55	1.75	1.32	1.30	1.17	1.72	1.88	1.21 (2)	1.76
DIVORCES AND REMARRIAGES									
(Div.or separate./married) · 1,000, 15-64	13.6	14.2	10.5	4.1	4.2	11.7	0.0	2.7 (2)	9.5
(Mar.>1° order/div.+wid.) · 1,000, 15-64	34.1	37.4	34.1	17.2	11.8	21.5	4.0	9.4 (2)	24.1

	NL	A	P	FIN	SW	UK	ICE	NO	CH
NUPTIALITY									
Mean age at first marriage for women	26.7	26.3	24.9	27.3	28.9	26.7	28.8	27.7	27.3
(Unmarried / women) · 1,000, age 35-44	131.7	123.6	86.6	192.6	255.0	107.9	208.1	153.9	136.6
Total first marriage rate · 100	56	55	73	59	44	53	55	58	64
FERTILITY									
(Extram. births / unmarried) · 1,000, 15-44	20.6	33.1	20.7	40.8	51.5	49.1	74.8	61.7	9.3
(Legitimate births / married) · 1,000, 15-44	93.4	71.5	76.8	88.5	74.0	84.6	76.8	82.2	97.5
(Births / women) · 1,000, 15-19	4.1	11.7	16.0	6.9	5.5	23.3	16.4	9.8	4.0
(Births / women) · 1,000, 35-44	27.7	17.4	19.0	28.5	26.1	25.9	37.9	28.2	24.3
Mean age at first birth	28.6	25.9	25.9	27.2	26.8	27.4	25.0	26.5	28.6
TFR of women born in 1955	1.87	1.77	1.97	1.88	2.03	2.02	2.47	2.05	1.75
TFR of the year	1.53	1.42	1.43	1.76	1.60	1.72	2.12	1.89	1.50
DIVORCES AND REMARRIAGES									
(Div. or separat. / married) · 1,000, 15-64	11.6	11.5	6.3	15.9	16.5	16.0	13.2	13.7	11.5
(Mar.>1° order / div. + wid.) · 1,000, 15-64	28.8	29.4	17.7	21.4	19.5	35.4	29.6	27.4	29.9

B: Belgium; DK: Denmark; G: Germany; GR: Greece; SP: Spain; F: France; IRE: Ireland; I: Italy; L: Luxembourg; NL: The Netherlands; A: Austria; P: Portugal; FIN: Finland; SW: Sweden; UK: United Kingdom; ICE: Iceland; NO: Norway; CH: Switzerland.

Source: Cantisani and Dalla Zuanna, 1999.

Figure 1. Association between TFR, extramarital fertility and marital fertility. 18 western European countries in 1981, 1991 and 1996.

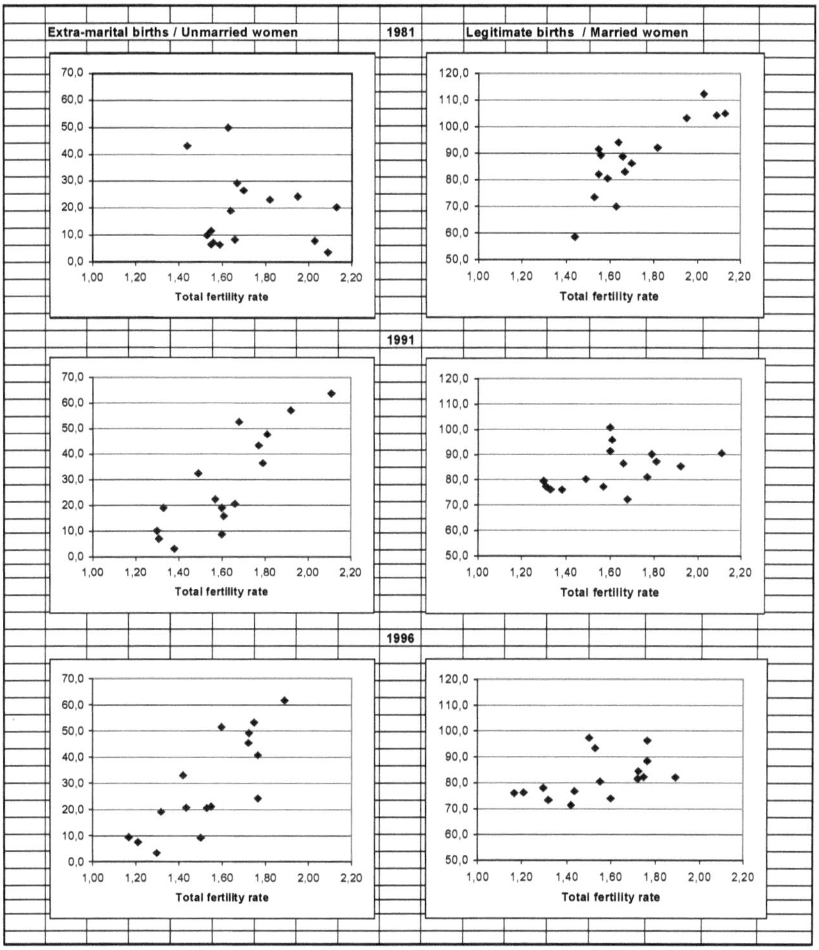

Source: Cantisani and Dalla Zuanna, 1999.

A familistic interpretation of marital and reproductive behaviour might be useful in explaining the peculiarities of other countries, such as Spain and – with marked differences – Japan, as some comparative research could show (Billari *et al.*, 2000; Dalla Zuanna *et al.*, 1998). Nevertheless, before extending the hypotheses discussed here to other countries and populations, systematic comparisons should be developed.

2. THE FAMILISTIC ITALIAN WAY OF LIFE

In chapter 2, Reher pointed out how the NCEC and Mediterranean countries are characterised by two different family systems: the weak and the strong family, respectively. In the weak-family area family ties (between parents and children, and among siblings) become less and less important during the teens and after the second decade of life. It is an ingrained anthropological feature, has by no means waned, as shown also by Kuijsten (1996), when studying recent social and demographic behaviours. Following in the footsteps of Reher, in chapter 3 Micheli underlines that in Italy as a whole the anthropological family structure is largely based on kinship, even if the rule of residence of the couple (patrilocal *vs.* neolocal), the diffusion of the stem-family, and the age at first marriage of both spouses show deep differences among counties. The Reher's classification of European families (strong *vs.* weak family) is reinforced by Micheli's analysis, who places the variegated Italian context in a European framework. The connection between family ties (strong *vs.* weak) and reproductive behaviour has not been developed by either Reher or Micheli. Nevertheless, as during the last 20 years in the strong-family area fertility has been the lowest world-wide, the social rules underpinning the strong-family system may be similar to those underlying the lowest low fertility. Consequently, a better definition of these could be useful.

Much of this task has already been done by various anthropologists, sociologists, historians and demographers, who have concurred in describing the familistic way of life (Banfield, 1958; Aldmon and Verba, 1963; Balbo, 1976; Ginsborg, 1989, 1994, 1998; Saraceno, 1994; Dalla Zuanna, 1995).

Following these authors, in a familistic oriented society, most people:

1. consider their own utility and family utility as being one and the same thing;
2. believe that every one else does too;
3. follow these two rules throughout their lives.

These rules especially concern the relationships between parents and children; having guaranteed his/her nuclear family, each person then extends these rules to the kinship. In other words, in a familistic oriented society, throughout their life most people seek their own happiness and at the same time that of their nuclear family and – if possible – their relatives. This familistic way of life can be considered as the "spirit" of the strong-family system described by Reher.

Table 2. Standardised □regression coefficients between total fertility and its three most proximate components (marital fertility, proportion of married women, extramarital fertility). 17 countries of Western Europe

	Marital fertility	Proportion of married women	Extramarital Fertility	R^2
1881	1.00	1.06	0.18	0.84
1911	0.85	0.53	0.19	0.99
1921	0.83	0.27	0.23	0.96
1931	0.90	0.38	0.18	0.99
1961	1.53	0.93	0.27	0.99
1971	0.92	0.01	0.25	0.78
1981	1.05	0.17	0.44	0.96
1991	0.58	0.05	0.89	0.94
1996	0.48	0.02	0.89	0.91

Source: For the period 1881-1971 the source is Coale and Cotts Watkins (1986, pp. 78-152). As indicators, I_f, I_g, I_m, I_h, calculated for the Princeton project were used. France was excluded from the elaboration, being defined as an outlier due to its role as precursor in the fertility decline. Correlation values, particularly for the first two years, would have been largely determined by France if it had been included in the regression analysis. Two separate values were given for England-Wales and Scotland. For the final three years (which included France) the following indicators were used: TFR for total fertility, *(legitimate births / married women)* $_{15\text{-}44}$ for legitimate fertility, *(unmarried women / women)* $_{15\text{-}44}$ for the number of single women, *(extramarital births / unmarried women)*$_{15\text{-}44}$ for extramarital fertility. As there is no large difference in the age structure of the total population and unmarried and married women in the 18 countries, it was not necessary to construct the Princeton indicators for the last two years. For further details, see Cantisani and Dalla Zuanna (1999).

Further to this definition, familism is not a general attitude toward the "traditional family", based on marriage and children, with the bread-winner father and the housewife, also found in demographic literature (Lesthaeghe and Meekers, 1986; Krishnan, 1990). As described in section five of this chapter, familism and the traditional family are linked, but familism can persist even where traditional family-life declines.

This emphasis on familism in interpreting Italian society has not been readily accepted universally. Some authors radically criticise familism as being an explanation for certain characteristic features of Italian society (De Masi, 1976; Gribaudi, 1994). Familism is not considered as a cause, but rather as being the effect of poverty and underdevelopment (in the past) and of the incapacity of the State to guarantee impartial welfare and rights to its citizens (today). Moreover, it is stated that it is not possible to speak of "Italian families" without considering the differences between north and south, town and country, and social class (Gribaudi, 1994). In answer to these criticisms, it is difficult to fully subscribe to the viewpoint of those authors who overemphasise the importance of familism in Italian history and society (e.g., Altan, 1986). Nevertheless, abundant empirical evidence shows that currently in Italian society – not only in its backward and archaic strata – the familistic viewpoint should be considered to better explain many kinds of social behaviour. Furthermore, Reher's considerations on the historical roots and the persistence throughout the centuries of the strong-family system in Mediterranean Europe suggest that parents transmit the familistic way of life to their children, an "explanatory variable" rather than the consequence of other social processes.

The authors mentioned above have studied the influence of familism on several sectors of Italian life, implicitly assuming that the intimate structure of family ties orients various social organisations: political parties, trade unions, businesses, universities, criminal groups and so on. Many authors underline the strength of the family as an institution contrasting with others – church, government, the community, *etc.* (e.g., Ginsborg, 1989, 1998). Moreover, it is difficult to develop this macro-social perspective without considering the micro-level of familism, which is rooted in the psychological and anthropological rules described above.

Let us consider some general data connected with Italian familism, comparing Italy with other Western countries. These general topics will be of relevance below, when the late departure from the parental home and lowest low fertility are considered.

Trust in others outside the kin is considerably lower in Italy than in any other Western country (Table 3). This is a possible indicator of the Italian familistic way of life, with trust in the kinship being very high (see Table 9 below). This dramatic difference is an important key in understanding the low esteem with which civic values are held in Italy. As the aforementioned authors found, familism and civic values cannot easily walk hand in hand: in a familistic oriented society, the saying *homo homini lupus* (every man is a wolf for the other man) should be substituted by *familia familiae lupa* (every kin is a wolf for the other kin) – without offence to wolves!

Another important consequence of Italian familism – and the low esteem of civic values – is the gap between private wealth and the quality of public services (Saraceno, 1994). In Italy, public services are – generally speaking – at a low level, if *per capita* income is considered. Perhaps the most clamorous example is the housing policy, where tenure has been positively encouraged, nor can public housing policy be compared to the majority of the NCEC. Thus the anxiety of the part of Italians to become house owners, as can be seen in Table 4 for the period 1961-91, which has continued throughout the 90s, with no sign of a let-up. This hampers any decision on the part of young people to leave the parental home, cohabit and marry. Moreover, given these trends on the housing market it is not easy to change residence and to do so a thorough knowledge of the local area is needed to find a dwelling. Many prospective house owners are actively encouraged to live near their family, although this is often more a question of choice, in keeping with the prevailing familism, rather than necessity (Table 5).

This could lead to the belief that in Italy welfare expenditure is low, which is simply not the case. The problem is that most welfare expenditure is directly transferred from the State to families, rather than to the public services. Moreover, most of these transfers go to the aged, financing pensions, whereas young people and couples with children are overlooked. This topic, which is very important in discouraging fertility, will be reconsidered below.

Table 3. Percentage of people declaring that the most of their compatriots are "very" or "somewhat" trustworthy. Year 1986.

Italy	GR	P	SP	F	IRL	B	UK	NL	G	LUX	DK
33	49	62	68	69	72	76	79	80	84	85	88

Source: Eurobarometro, Bollettino Doxa, XL, 22-23, Nov. 17th 1986. Data quoted by Inglehart (1990). For the abbreviations of the countries, see the note of Table 1.

Table 4. Percentage of households owner of their houses.

Italy				Other Europeans countries (1990-91)							
1961	1971	1981	1991	GR	SP	F	B	UK	NL	G	DK
46	51	59	68	76	78	54	54	67	45	39	54

Source: For Italy: census data; for other countries: census data collated by Eurostat. For the abbreviations of the countries, see the note of Table 1.

Table 5. Residence of parents and children in some industrialised countries during the 1980s.

	UK	USA	AUS	G	A	HUN	Italy
Proportion % of parents living with at least ...							
... an adult son	32	21	30	40	39	37	60
... an adult daughter	29	14	25	26	25	30	58
Proportion % of adult children not living with parents, whose mother living							
... 15 minutes of less	32	27	24	38	37	43	57
... between 15 minutes and 1 hour	40	31	33	30	35	35	26
... between 1 and 5 hours	19	19	20	22	23	19	8
... 5 hours or more	9	23	23	9	4	4	4
% of adult people living near their mother (1 hour or less) who see her every day	11	16	7	20	17	32	32

Source: Jowell *et al.*, 1989, quoted by Ginsborg (1994). For the abbreviations of the countries, see the note of Table 1.

The Italian economy is also deeply steeped in familism. The production sector is characterised by thousands of small firms, whose founders are usually siblings or other relatives (see Table 6). Many years after the company's founding, owners and managers often continue to belong to the same family as the founder(s) (children, their spouses, nieces, nephews, grandchildren). This kind of organisation is not only typical of small firms: the Agnelli dynasty and the four Benetton brothers are but two examples of the success of this kind of model on broad industrial dimensions.

Table 6. Some data on dimension of firms. Some EU countries around 1995.

	Italy	SP	F	UK	NL	G	SW
% industrial income produced by firms with fewer then 50 workers (1)	51	42	39	28	41	31	33
% industrial workers employed in firms with fewer then 50 workers (2)	64	63	50	43	42	41	43

Source: Eurostat Yearbook, 1997 (Italian version), p. 366. For the abbreviations of the countries, see the note of Table 1.

Moreover, the Italian industrial system is strongly oriented towards the production of consumer goods that emphasise the quality of life, particularly inside the home (see Table 7). This last topic will be reconsidered below when considering the impact of increased relative economic deprivation from a familistic perspective on fertility decline.

Table 7. The "house quality " industrial sectors where the Italian trade balance was the best in the world (millions of $). Year 1996.

Products	Italy's commercial trade positive balance	Italy's principal competitors (commercial trade balance in brackets)
Ornamental stone	1,702	China (100)
Ceramic tiles	3,327	Spain (1,427)
Chairs and sofas	3,126	China (421)
Furnishings and kitchens	5,023	Denmark (1,350)
Lighting and illumination	1,168	China (1,081)
Taps and valves	2,378	Germany (1,815)
Marble cutting machinery	682	Japan (297)
Locks and ironmongery	1,054	Taiwan (526)
Stove and hobs	506	China (107)
Radiators and boilers	614	Germany (291)
Fridges and freezers	2,431	The USA (1,465)
Washing machines	1,701	Germany (327)

Source: Fortis, 1998.

Finally, both sides of the coin are evident when considering the labour market. Thanks to family assistance, Italian society can bear high youth unemployment rates (more than 50% in some southern provinces), in the absence of public unemployment benefit and social upheaval. But the influence of the family on the labour market is even greater. More than 30% of young workers – interviewed in 1996 – found their first job thanks to the direct intervention of a relative, 50% under their own steam – although some contribution from the family is to be suspected – whereas less than 20% utilised other resources (advertisements, public agencies, *etc.*) – Buzzi *et al.*, 1997, p. 372. While the role on the part of the family ensures greater flexibility for those entering the labour market, on the other hand many young people are not actively encouraged to seek employment. It is often preferred for the young to stay at home, unemployed, rather than accept a low status occupation. As will be seen parents find it hard to accept a "low status" child. Consequently, in many parts of unemployed southern Italy, during the last years immigration from developing countries has been relatively high and growing.

3. FAMILISM AS AN OBSTACLE TO LEAVING THE PARENTAL HOME

Castiglioni and Dalla Zuanna (1994), in a previous analysis of a 1991 national survey, described the parental family as the golden cage of Italian youth. Later surveys in 1995 (Buzzi *et al.*, 1997) and 1998 (IRP, 1999), the latter with a more demographic focus, showed that there was a trend in favour of greater autonomy but that this continued to be based within the family home (Table 8). Familistic parents do not encourage their children to leave, and young people react by placing their trust and affection in their parents and relatives (Table 9).

Reher and Micheli suggest that this situation is rooted in the past, *i.e.* in inter-generation cultural transmission. Perhaps, it would be useful to explain this situation from the behavioural viewpoint, using the familistic paradigm. Let us first consider the direct effects of familism on the late leaving the parental home. Usually, children living outside the family pay for their freedom with a loss of amenities (Hill and Hill, 1976). But Italian parents are reluctant to see their children suffer in material terms, in the belief that their offspring's discomfort is a source of their own malaise. Therefore, parents discourage an early departure from the family home and anyone who does so has to be strongly motivated (*e.g.* burdened by strict rules and limited freedom). Nonetheless, the golden cage is usually very comfortable and almost invisible.

Also directly connected with the late departure from the parental home is the parents' major emotional investment in their children. In the aforementioned IRP survey, both adult children and parents stated that the main disadvantage in leaving the home was the emotional price paid, and as many as 60% of the parents interviewed said that the child's departure offered no advantage of any kind.

Table 8. Some data on living arrangements and earnings of Italian people aged 15-24, interviewed in the 4 rounds of IARD survey

	1983	1987	1992	1995
Percentage of people living outside the parental family	6	6	5	4
Percentage of young workers living at parental home who give all their earnings to the parental family	23	15	9	5

Source: Buzzi *et al.*, 1997.

Table 9. Mean score assigned to cohabiting relatives by Italian people aged 20-34 in 1998. Scale from ONE (the lowest) to TEN (the highest).

Father	Mother	Brothers	Sisters	Grand-parents	Other relatives
7.84	8.55	7.99	8.17	8.26	8.13

Source: IRP, 1999.

This co-operative adult family (I thank Rossella Palomba, co-ordinator of the IRP survey, for this felicitous terminology) is based on a double exchange: both parties (parents and adult children) give and receive material and emotional goods: parents give more material goods and receive more affection, adult children give more affection and receive more material goods. For some authors, this material exchange, at the end of the life course, will be more balanced in favour of parents, if the pension system is considered (Cigno and Rosati, 1992). Nevertheless, it is less risky, for adult children, to stay on in the parental home, than to risk building a new family, where affective and material goods are re-contracted, on a daily basis, with their partners.

Some might say that this co-operative family of parents and adult children is the modern version of the court of Aeolus – described by Homer at the beginning of the 10th Book of the Odyssey – where the God of the winds gives his six daughters as wives to his six sons, and offers the six couples an eternal banquet (it is also an extreme example of endogamy, and it could be a good preventive measure against the Oedipus murder!). In Italy, few young people are as bold as Ulysses, leaving the banquet provided by Aeolus to face the open seas.

A familistic oriented society also discourages early departure from the parental family for a number of indirect motives, too. As suggested above, familism has contributed to the shortage of housing for rent accessible to young people. Moreover, familism was an important brick in building the Italian welfare system, largely based on the private transfer to old people rather than recipes from the public sector and the payment of unemployment benefit for all. Nevertheless, any material obstacle to leaving the parental home should not be over emphasised, as even the young employed with a good income or home owners usually prefer to remain until marriage (to the joy of their parents!).

Before concluding this section, it could be useful to reconsider the Italian situation using the household formation framework proposed by Burch and Matthews (1987). Following these authors, the answer to the question "With whom shall I live?" stems from the necessity to procure household goods, which are both material (*e.g.* domestic services) and immaterial (*e.g.* companionship and privacy). The value (the cost) of these goods is not fixed, but varies with the evolution of the society, and – for the decision maker – can be considered as exogenous. The researcher could find an answer to the above question defining and measuring this utility function. Looking at "the modern Western society", Burch and Matthews define some explanatory hypotheses (*i.e.* the utility function), in order to explain the growing proportion of people living alone (or in a small household). Trying to fit Burch and Matthews' hypotheses to the living arrangements of Italian young people, we discover that they do not refer to "the modern Western society", but "the modern weak-family Western society". The more clamorous example concerns the effect of the rising real income. For our authors: "We are on safe grounds in assuming that one reason so many people now live alone or in very small households is that they can afford to. They are able to forgo the economies of scale represented by larger households…"(p. 503). This could describe some aspects of the Italian situation, but is not useful in understanding why adult children stay in the parental home until their thirties. In Italy, the increasing income of familistic parents increases the amount of money available to their adult children. Empirical data show that during the last 20 years, the proportion of young people in employment living in the parental home, with their salary at their complete disposal, has substantially increased. At the same time, for those contributing to the household expenditures, the proportion of salary given to their parents has substantially dropped (Buzzi *et al.*, 1997). This situation – that is strictly related to the familistic rules described above – increases the opportunity-cost of leaving the parental home. Hence, in a familistic oriented society, the rising of real income hampers – rather than favours – the early departure of children.

Another behavioural hypothesis described by Burch and Matthews concerns the rising demand for privacy: "With higher real income and a sense of security provided by an extensive welfare net, the individual turns inward and becomes more concerned with self-development and personal growth and experience. Recipes for such growth often emphasise the need for solitude and privacy" (p. 505). In Italy, the growing demand for privacy coming from adult children is often resolved in the parental home. Data show that during the last 20 years a growing proportion of adult children (both males and females) have a room of their own in the parental home (77% for people aged 25-34 in 1998), where they are often free to pursue a sexual relationship with their partner (57% for people aged 20-34 in 1998) – Buzzi *et al.*, 1997; IRP, 1999.

Summing up, Burch and Matthews' framework is very stimulating and the explanatory hypotheses could hold also for Italy, referring to the living arrangements of older people and other demographic behaviour reducing the dimension of the household. But the explanatory hypotheses should be rearranged to understand the long permanence of adult children in the parental home. The familistic bonds between parents and adult children should be taken into account, as they define the milieu where the decisions concerning living arrangements are taken.

4. FROM THE LATE DEPARTURE OF THE PARENTAL HOME TO LOW FERTILITY

Many authors have described the psychological and practical consequences of remaining in the parental home on several aspects of Italian social life and private behaviour – see chapter 3 of Ginsborg (1998) for a review. The direct and indirect consequences of the late departure from the parental home on fertility are stressed in the following section, which could help cast light on Figure 1.

The direct effect is not as important as it might seem. In Italy, as elsewhere, age at leaving the parental home is negatively associated with the *quantum* of fertility (Billari *et al.*, 1999a, 1999b). Nevertheless (Table 1), in 1996 in Italy the mean age of women having their first child is not particularly high: 27.8 years, as in the UK and France, where early departure from the parental home and cohabitation are widely diffused. In other words – looking at large populations – it is possible to have a late age at first child and a TFR not far from the replacement level, even if (looking at individual behaviour) the later the age at first child, the lower the probability of having other children. Consequently, today, within a European perspective, the problem of Italian fertility is not its late beginning, but the low probability of having a third, second and – for the youngest cohorts – a first child.

Table 10. Percentage of Italian women without birth at the end of reproductive life by cohort

	1950	1955	1960	1961	1962	1963	1964	1965	1966
Italy	11	11	15	16	17	18	19	21	23
North	10	13	18	19	21	22	24	25	27
Centre	7	10	14	15	17	18	20	22	23
South	15	13	12	13	14	14	15	17	18

Source: ISTAT, 1998, diskette enclosed. Data until 1995 are calculated, after 1995 are estimated.

The main direct effect on fertility is that the late departure from the parental home often results in the failure to marry and have children. The number of childless women has strikingly increased among cohorts born after 1955, particularly in northern Italy. Table 10 is probably based on too low estimations of late fertility, and the proportion of those born after 1960 having at least one child will probably be higher. Nevertheless, childless Italian women born in the mid 1960s will surely top 20%: the proportion of childless women among cohorts born just ten years previously is going to be doubled. In this case, a delay of marriage often becomes permanent celibacy (De Sandre *et al.*, 1997). This is an indication of something amiss, *i.e.* an unsatisfied desire, since throughout the 1980s and 1990s all the surveys conducted very few young people expressed the wish not to marry or have children (see Table 11 below).

Let us consider the indirect effects on fertility of a late departure from the parental home. First of all, men have no experience of housework, since they go directly from their mother's arms to their wives, never having lived alone or with friends. Thus Italian husbands do not help out in the home, even if their wives are in full-time employment (Bimbi, La Mendola, 1999). The excessive burden for women can be considered as an important cause of Italy's lowest low fertility.

Another important indirect effect is less easily described and empirically supported. Staying at home until their thirties, young Italian people risk overemphasising each transition in psychological terms. Thanks to the cover offered by the parental family, they accept a job only if it is in line with their desires, and postpone marriage until the risk of losing amenities is low. In other words, during the age interval 20-30, when the enthusiasm for innovation should be higher, they fail to develop a taste for responsibility, almost indispensable to the transition to adulthood. Massimo Livi Bacci (1997) refers to this Italian situation as being *la sindrome del ritardo* – "the delay syndrome". Young Italian people often become precociously fervent supporters of Malthus. Garelli (1984) said that this insecurity among young people in facing transition is the product of the general level of insecurity present in society. Micheli (1996) proposed that the inability of today's children in making choices is the direct product of their parents' insecurity. Golini (1997) supposed that the modernisation of Italy was too rapid to permit the growth of balanced relations between generations. All these hypotheses are interesting in putting together the pieces of the puzzle underlying the motivations for the late departure from the parental home and Italy's lowest low fertility, but they are not enough to explain the broad differences between Italy and the NCEC. To better understand the intricacies of the Italian context, the familistic paradigm should be taken into account.

A brief summary could be useful. Using data from several sources, the connection in Italy between familism, late leaving the parental home and fertility, has been described for the last 20 years.

Familism encourages late departure from the parental family directly, because:

- In the Italian familistic society, economic conditions at home are more favourable than living alone, with friends or a partner;
- Generally speaking, in Italy the affective bonds between parents and children, children and other members of the kin are very strong.

Indirectly, because:

- Familism has contributed to the shortage of housing for rent accessible to young people;
- Familism was an important brick in building the Italian welfare system, based on private transfers from older to younger people rather than the availability of unemployment benefit for all.

Late departure from the parental home negatively influences fertility directly, because:

- The higher the age at marriage or cohabitation, the shorter the time-interval available for childbearing;
- Often the delayed departure implies a definitive no to cohabitation, marriage and childbearing.

Indirectly, because:

- Young men do not learn to do housework, and thus in the Italian couple the working married (or cohabiting) woman has a double role (in the market and *ménage*), without being helped by the husband;
- Staying on in the parental home until their thirties, young people forget how to risk, useful to tackle the prospect of childbearing (the "delay syndrome").

Up to now, the interaction between familism, late departure from the parental home and low fertility has been described. The puzzle has to be completed by discussing the effect of familism on the reproductive behaviour of Italian couples.

5. FAMILISM AS THE LUBRICATING FACTOR IN THE COUPLES' LOWEST LOW FERTILITY

To explain the falling fertility rate in Western countries from 1970-2000, Lesthaeghe (1998, 1999) emphasised three compatible theories: (1) increased female autonomy and opportunity-costs, (2) increase in relative economic deprivation, (3) changed cultural attitudes toward post-materialistic values. Let us examine if and how these interpretations can hold water in the familistic Italian society.

Following the first theory, better educated women has led to increased opportunity costs for them and, therefore, to a higher proportion of working women, to lower fertility and delayed marriage and parenthood (Becker, 1981). Several authors try to fit this theory in with Italian data, and an apparent statistical "paradox" has been found, when comparing individual and ecological statistical analyses. Fertility among more educated Italian women and those in employment is later and slightly lower than fertility among housewives (see, e.g., Rosina, 1999; Di Giulio *et al.*, 2000). Nevertheless, when Western countries are considered as statistical units, a strong positive association is detected between TFR and the proportion of women in the labour market (Pinnelli, 1995), Italy being characterised by both low fertility and low female employment. Similar results are obtained using multilevel models – pooling the FFS data of several countries – the first level units being women or couples, the second level units being the countries (De Rose and Racioppi, 2001).

To understand this "unusual" result, let us consider the fact that an Italian woman, who does not stop working after having children, must face at least four more "familistic" problems than NCEC mothers.

First of all – as described above – the Italian mother finds society organised as though the male-breadwinner family were dominant, since public child rearing services and nursery facilities are scarce.

Secondly, she and her partner receive no State support to face the expenses incurred by a new child, since the family allowance paid to families with children is extremely low (Ditch *et al.*, 1996). Let us only consider the fact that in 1994, 8 billion U.S. dollars were gathered as taxes to finance family allowances; only 2.5 billion were effectively re-distributed to families with children, whereas 5.5 billion were used to pay pensions (Saraceno, 1998, p. 104).

Moreover – as described above – the woman is rarely helped by her partner in the *ménage*: the time spent working (at home and outside) is dramatically higher for a working Italian married woman than for an Italian married housewife (Bimbi and La Mendola, 1999).

Last but not least, the Italian mother must face the anxiety of having very little time to spend with her child: she violates the well entrenched social norm – strictly connected with the familistic rules – that nothing is better for a child than to be with her/his mother.

Consequently, two main familistic influences may be detected in Becker's behavioural chain theory. Following a new-home economics' viewpoint, the Italian familistic organisation of couple and society

1. reduces the number of children to couples where both partners are earners, stressing the economic and psychological contrast between the woman's employment and childbearing;
2. induces Italian mothers into stopping work whilst the children are growing up, which is why Italy is characterised by low female participation in the labour market, that also applies to the northern regions where unemployment is practically unknown (Bettio, Villa, 1998). Moreover, half the Italian married FFS women aged 20-49 in 1996 were housewives at the time of the interview (Bernardi, 1999).

Social process (1) strengthens the negative association between female employment and fertility in Italian society. Thanks to social process (2), familism obstructs female employment: consequently, familism helps reduce both TFR and the proportion of Italian women inside the labour market.

Following the theory of increased relative economic deprivation, rising consumption expectations lead to the increasing value of private consumption. This process leads to competition between children and consumption, since high and rising consumption expectations can far better be satisfied by dual earner families (Easterlin, 1976). Italian FFS data seem to be consistent with these theoretical expectations (Di Giulio *et al.*, 2000).

This result are associated with the following familistic social processes:

1. As already described in the second section and in Table 7, Italians are strongly oriented toward consumer goods connected with the quality of life in the home. These consumer expectations delay age at marriage, and emphasise the cost of a second or third child; the parents must take into account new high quality furniture and a new high quality house;
2. When consumption aspirations rise, child value also increases, because familistic parents must compete with other familistic parents: adopting, at the same time, familistic and consumption oriented viewpoints, a child can be considered as being a luxury (De Santis, 1997). Puzzling over these considerations, material and immaterial investments made by familistic Italian parents in (what they consider to be) the quality of their children are rather high.

Three kinds of empirical data support these last statements. In Italy the cost of children is higher than in the NCEC, and this cost has been growing over the last few years (Ekart-Jaffé, 1994; De Santis and Righi, 1997). Moreover, *ceteris paribus*, children with fewer brothers and sisters have more possibilities of improving their own social class, thanks to better education (Casacchia and Dalla Zuanna, 1999). These results show that the strategy of reducing fertility has been a good familistic tool in Italy over the last 30 years, helping the social climb of few children or the only child.
The third empirical proof can be found in the survey data on the value of children. Table 11 illustrates the results of a survey conducted on two comparable samples, interviewed with the same questionnaire in Italy and in the Netherlands in 1989. The values assigned to children were clearly higher in Italy; from a familistic viewpoint, the interpretation of these results by Rossella Palomba are easily shared:

> The parent-child relationship has been reworked and adjustments have been made in modern times (...). Today the relationship offers different possibilities: an insurance against loneliness, finding a purpose in life, and acquiring a social identity. Italians are not in conflict with traditional values; they are merely adapting them to fit in with today's society and with changing times. In this way, there is not conflict between modernity and tradition as regards values and children, and Italians place a very high value on children. This high value could, paradoxically, explain the low Italian birth-rate, since Italians invest a great deal of time, money, attention, and interest in their children and thus, instead of having several children, they prefer to have only the one child, who becomes the sole object of much care and attention (Palomba, 1995, p. 186).

The third theory has been formulated and re-adjusted by Lesthaeghe, van de Kaa and others (Lesthaeghe and Meekers, 1986; van de Kaa, 1988; Lesthaeghe, 1995). The basic idea is that new European fertility and marital patterns cannot be interpreted without starting from changes in mentality. As data for Western countries show, cohort by cohort the orientation toward post-materialism increases, and this new pattern of values encourages cohabitation, low fertility and couple dissolution.

In Italy post-materialism is at a lower level than NCEC, even if new cohorts are fast making up for lost time (Lesthaeghe and Meekers, 1986; Inglehart, 1997; Ginsborg, 1998, pp. 244-245). Moreover, the traditional indicator of secularisation (the weekly participation at mass), after a rapid decrease during the period 1950-81, remained substantially stable around 30-35% until the mid-1990s (Pisati, 2000). Generally speaking, religion has been gaining popularity over the last twenty years, particularly amongst the young (Table 12), even if many authors and data suggest that Italians have reverted towards an intimate and self-tailored Catholicism, which is less oriented by Church teaching (Cesareo *et al.*, 1995).

Table 11. The values of having children in Italy and the Netherlands. Age group 20-44. Percentage who agrees with the statement.

	Italy	Netherland
You cannot be really happy if you do not have children	57	7
Having children is your duty towards society	43	8
The closest relation you can have with anyone is with your own child	87	38
You can be perfectly satisfied with life if you have been a good mother or father	79	51
In our modern world the only place where you can feel completely happy is at home with your children	50	32

Source: Moors, 1990.

Table 12. Answer to the question: "Is the religion important in your life?" Four rounds of the IARD survey. Italian people aged 15-24, column percentage.

	1983	1987	1992	1996
Very much	7	9	10	12
Much	20	22	22	23
Enough	37	38	37	33
Few	24	23	19	22
No	11	8	10	8
I don't know	1	0	2	2

Source: Buzzi *et al.*, 1997, p. 424.

Dealing with the empirical connection between fertility and post-materialism and secularisation indicators (the indicators of these two factors being strictly related – Clerici, 1999), the statistical "paradox" described above is found again. Among the more secularised Italian women and counties, fertility is later and lower (Clerici, 1999; De Sandre and Dalla Zuanna, 1999; Di Giulio *et al.*, 2000; Dalla Zuanna and Righi, 1999). Nevertheless, assuming a European viewpoint (where countries are considered as statistical units, adopting both ecological and multilevel analyses), fertility is lower in the less secularised countries – De Rose and Racioppi, 2001.

The social processes underlying this "paradox" could be similar to the ones already seen whilst discussing Becker's theory. The change of values toward individualism and secularisation could be obstructed by the underlying familistic mentality: consequently, in Italy both fertility and secularisation are low because they are both dropped by the Italian familistic way of life. The above sentences can be discussed in light of research conducted by some authors and their data. Firstly, familism cannot cohabit with an excessive individualism, because it is a sort of familiar individualism (each kin for the other kin is a wolf...) and during the 1980s and 1990s, after the decline of the collective dreams of the 1960s, familistic values have celebrated their Italic triumph (Ginsborg, 1993, pp. 557-566; 1998, p. 533). On the other hand, familism does not contrast with some Catholic values: *e.g.* the idea that families should be defended – rather than substituted – by the State; the emphasis on the responsibility of parents towards their children. These Catholic values are largely shared by Italians, while other non-familistic Catholic values are less and less popular, even amongst those closest to the Catholic Church (Clerici, 1999; Ginsborg, 1998, pp. 233-237). An interesting example is the result of the two referendums on divorce and abortion, carried out in Italy – respectively – in 1974 and 1981. One might expect that the votes cast against divorce were less than those cast against abortion, because abortion – generally speaking – is considered by the Catholic ethic as a murder, a greater wrong than divorce. Nevertheless, people voting against divorce were 44% whilst there were only 34% against abortion. An interval of seven years could be large enough in order to explain this difference, as they were years of major cultural changes, particularly in sexual and reproductive behaviour (Castiglioni and Dalla Zuanna, 1995). Nevertheless, another explanation should be added. At the time of the vote, many Italians may have adopted a familistic viewpoint, considering divorce as a threat to the family, while abortion was seen as a resource, *i.e.* the possibility of solving a private problem inside the family – without fuss, free of charge and with the help of the Public Health Service: *i panni sporchi si lavano in famiglia*: "do not wash your dirty linen in public ".

Summing up, two social processes are particularly relevant, to understand connections among familism, Catholicism and fertility in Italy:

- Catholic values are filtered by the familistic way of life. As the Catholic Church has emphasised some values easily compatible with familism in Italy, Catholicism has reinforced familism, and – partially – viceversa. This social pattern hampers secularisation;
- Some non-familistic Catholic values are less and less popular among Italians and their families. One of these is the non-familistic and un-

popular Catholic value of a large number of children, strongly recommended by the Pope and Italian bishops, but practised only by a small minority of Italian couples.

The consequent result is the "paradox" described above: the most traditional and Catholic Italian couples and counties have a higher fertility than the majority of secularised ones; but if Italy is compared with other European countries, a lower secularisation is associated with lower fertility.

6. FINAL REMARKS

The forces reducing fertility are not particularly different in Italy compared with the NCEC. But familism has interacted with them, emphasising the strength of the social processes, increasing the number of persons without children and depressing the fertility of the couple. The persistence of familistic rules is useful in explaining the lowest low fertility in Italy during the last 20 years, compared to the NCEC weak-family area.

Should the social process described here effectively be at play, Italian fertility is unlikely to perceptibly increase without undermining the strong-family system. But the secular anthropologic structure of the strong family and kinship is not easily slackened. In the Italian context this social-anthropologic system and the interconnected familistic norms hold, despite far-reaching economic, social and cultural change. Novelties have been assimilated into Italian society without loosening (and perhaps even enforcing) its basic familistic structure. Perhaps, even the diffusion of "new" marital and reproductive behaviour (divorce, cohabitation, extra-marital fertility) – that could apparently weaken the strong-family system – actually reinforce it. To give an example, in Italy – as elsewhere – the bonds between an adult child with his/her parents are often strengthened by a divorce. The adult child can find in his/her parental family the psychological and material supports to face the shock of a marital dissolution. Perhaps in the strong-family system – generally speaking – the weaker the ties between partners, the stronger the affective and material connections between adult children and their parents and kin.

Gianpiero Dalla Zuanna, Department of Statistical Sciences, University of Padova.

GIUSEPPE A. MICHELI

ON THE VERGE OF A FAMILISTIC INTERPRETATION

Familism, moods and other alchemies

1. BEFORE THE FIXED ENDPOINT: WHAT SET OFF THE CHANGES IN SOCIAL NORMS?

Some believe that what we are dealing with is a particularistic variant of familism, others that it is a sublimation of the latter, in the form of amoral individualism. However, there is wide agreement that we are now facing a situation in which the family is regaining a central role in social and economic reproduction. What is it that set off the changes, which have led from a scenario of amoral, and solidaristic familism, which lasted so long over time, to the present situation? When did it do so, and why? What explains the prolongation of the trend of the decline in family-oriented choices, over so long a period as to occasion a breaking away from previous models? More in general, if a new system of norms has been formed (favourable to new roles and affecting individual choices), when and how did this come about?

Let us attempt to get closer to these important questions, by starting off with a crucial issue emerging from the evidence of many surveys: the proper number of children declared by Italian women remain stable, around two per woman (Goldstein *et al.* 2003). A similar result emerges from a general survey conducted in the Friuli-Venezia Giulia region on a sample of women from three distinct cohorts.[1] The data of the Abacus Social Barometer also display extraordinary stability in the same proper number in all the regions of North Italy and for all the age groups.[2] The 1945-47 cohort (which reaches a peak in its reproductive life span in the mid-seventies) effectively fulfilled this objective, but for the subsequent cohorts, this goal seems further off. How can actual patterns of behaviour be reconciled with declared expectations?[3] And above all, how can these expectations be reconciled with the traditionally recognized mechanisms of their formation? Let us start with four considerations, based on the findings of the Friuli survey.

G. Dalla Zuanna and G.A. Micheli (eds.), *Strong Family and Low Fertility: A Paradox?*, 127-160.

The first consideration concerns the divergence between declared preferences and personally experienced costs. Reasoning strictly according to the logic of rational choice, those who experience costs and sacrifices should be induced to scale down the goal regarded as ideal for a family such as their own, and subsequently all associated behaviour. This appears to be true for the older cohort of respondents, but not for the other two younger cohorts; for the latter, the experiencing of costs and sacrifices at the birth of the first child does not lead to a reduction in the family size regarded as desirable.

Secondly, this finding may confirm the hypothesis that anticipatory socialization to family-model norms may have a decisive influence on declared preferences. In effect, in the two younger cohorts of the Friuli survey, the proper number of children appears to be influenced by the size of the family of origin: it is children from large families who aspire to having the most children themselves (as one might expect). But this rule does not hold for the oldest cohort. The powerful mechanism of socialization breaks down in the very case of the cohort born in the years immediately after WW2.[4]

Let us suppose, thirdly, that the normative framework influences the formation of life projects. Let us consider the average proper age for the first child: this rises gradually, from 25 to 27 years, as we pass from the oldest cohort to the younger ones, who have experienced less of the costs and sacrifices inherent in motherhood. What is it that justifies a change in the norms? The delay of timing among the younger women might not depend so much on personal experience of the cost of these transitions as on the spreading of this information through the network of friends and acquaintances. But if the intentions and projects of individuals are influenced by a widespread system of beliefs, this influence should hold strong for all women exposed to this process of social learning. On the contrary in the two younger cohorts, the putting off of the proper age for parenting decisions appears to be significantly higher among those respondents who have not experienced any sacrifice or cost. In short, there does not appear to be any confirmation of the hypothesis that a changed framework of costs (individual beliefs) produces a changed framework of individual preferences as regards the timing of life events. Rather, it would appear that a different shift in perceptions, i.e., an increased intolerance of the costs of possible choices, might be responsible for the shift in preferences. The question therefore remains unresolved (undermining any rigidly cognitivist approach): what is it that produces the change in norms and collective tastes?

Figure 1. A first explanatory model of inconsistency between behaviour and expectations: Rational choices vs. familistic framework.

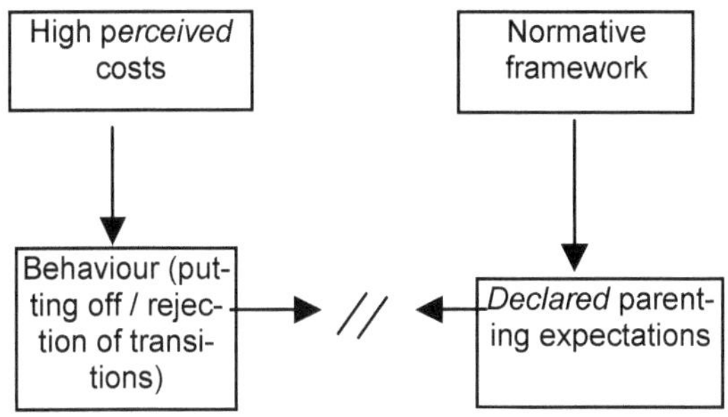

Elster (1989) observes that the fox that is after the grapes, but for some reason cannot reach them, will tend to diminish their value and change either its belief system ("the grapes are sour") or its system of preferences ("I don't like them anyway"). Let us suppose that having a child remains a desirable goal for the younger cohorts, but that for some reason this becomes difficult to achieve. The reduction in the dissonance between these two cognitive elements may be obtained either by changing the system of beliefs (e.g., by overestimating the costs of achievement) or by altering some attribute of the goal desired, so as to decrease its desirability (the irreversible nature of choices associated with the formation of a family has often been read, in processes of modernization and individualization, as an infinite cost[5]). We would therefore expect there to be some degree of consistency between the perception of high costs and the dismantling of the reversibility of traditional irreversible transitions. But in the Friuli sample, it is just those younger cohorts with an accentuated perception of the costliness of family formation who return to a mass acceptance of the idea of the irreversibility of such transitions. The model of family formation for the young Friuli cohorts respects traditional canons: it is both desirable and irreversible. But at the same time it is too costly and so tends to be postponed. Among the younger cohorts, although the normative framework of the traditional family model still prevails in the formation of decisions, the perception of the costs of family passages is greatly increased. This leads to a clear-cut bifurcation between declared expectations (socially ruled) and rational demographic choices (Figure 1): the dissonance between expectations and reality would be open and lacerating, and there would be no attempt underway to reduce it.

The scarce predictive ability of attitudes measured in this case is not an isolated fact, and is deserving of a brief reflection. Opinion surveys traditionally sketch out a picture of the decision-making process which reads basically as follows: it is a mixture of social background, socio-cultural climate and the set of individual and collective values which influences the individual's attitudes, and these in their turn anticipate the corresponding choices. According to this picture, attitude is a mere theoretical construct of socio-psychological models.[6] But in the studies on the value of a child, the value-expectation models were shown to be "inconsistent and inconclusive" (Crosbie, 1984): if any merit can be attributed to them, it consists precisely in having ascertained that there is no clear functional link between the declared system of preferences and values and behaviour, either concomitant or both potential and subsequent in time. The low predictive value of attitudes derives from a set of reasons of varying importance. The most radical criticism[7] (Festinger, 1964) overturns the traditional sequence, whereby attitude precedes action, both chronologically and genetically: attitude is, in contrast, a *post factum* processing of actual behaviour. It has the hidden objective of giving structure and legitimacy to past or current behaviour, not future behaviour. Behaviour therefore finds its breeding ground not so much in the cognitive sphere, as in the normative-affective sphere: behind individuals' rational choices, there is a substrate of dispositions which affect them (without actually determining them) and which operates not at the conscious level of beliefs and preferences, but at the more basic one of desires and experiences.

How then may we interpret the persistent discrepancy between norms, values and attitudes, on the one hand, and behaviour on the other, to be found among the women in the Friuli survey? A first interpretation is based upon the hypothesis that there is no dissonance between an increased perception of the costs of transitions and a persistence of the familistic model, but that – in contrast – it is the very accentuation of a familistic vision of the family that reduces its size, increases its costs and lengthens the intervals between births. This is the interpretative key proposed here by Dalla Zuanna, which is consistent with the observation of the persistence in Mediterranean Europe of anthropological models based on the strong family (Reher) or on the stem-family (Micheli, first contribution).

It is, however, worth following a second line of interpretation in order to explain why among the younger cohorts "the grapes are out of reach, but they are not sour", i.e., why there is an increased perception of the costs of family passages, but this does not affect the perception of their irreversibility. This reading justifies the inconsistency between the set of internalised norms[8] (effectively oriented towards traditional models of transition to the adulthood) and consequent behaviour, without having to turn to the intermediate category of attitude, but rather by positing the existence of a mechanism of interposition in the expectation-behaviour sequence, an interceptor mechanism[9] which cannot be identified in the cognitive constructs generally taken into account (Figure 2). It is as if antagonist mediators interposed themselves between the decisions and stratification of awareness upon which the system of preferences and expectations is based, as happens in the processes of endocrine communication and cerebral synapses.[10] These direct the individual towards this or that reaction on the basis of the way in which the individual's life events combine with his dispositions; alternatively, they may even intercept every reaction, rendering the individual apparently lacking in reactivity, for no apparent reason.

Figure 2. A second explanatory model of inconsistency between behaviour and expectations: Post factum rationalization of intercepted choices.

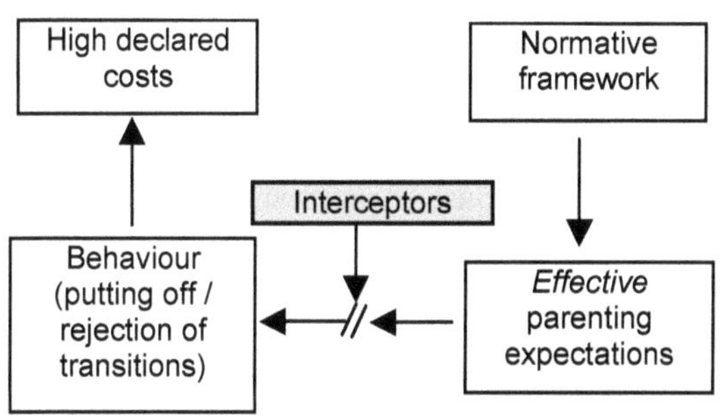

There are therefore two distinct models of interpretation of the inconsistencies between behaviour and expectations. According to the first, it is the framework of a familist philosophy that is decisive; according to the second, it is an 'irrational' tendency to avoid the irreversible. Both are legitimate. Indeed, if a change in individually absorbed norms can descend directly from changes in the framework of individual costs and benefits, it will also be able to descend from a change in the collective framework of symbolic references, emotional drives of attraction or repulsion. These changes are certainly located within a black box (Boudon, 1998), which it is difficult to identify and describe in terms of individual rationality,[11] but the explanation of the change in norms and preferences may then lie precisely in these important social facts.

Let us therefore pose a few questions. When norms do not produce a manipulation of practices, but absorb the manipulation produced by the latter, what is it that produces the change in the practices themselves? If we are to explain the demographic choices declared by individuals, we have at hand the two usual models of synchronic mechanisms of decision making: the economic one of cost-benefit analysis, and the sociological one of acceptance of norms and sanctions. Are these really enough? Or should we also investigate the subjective formative frameworks of decision-making processes?[12]

> "Two strategies of research deserve our attention. The first, at a micro-demographic level, should pursue the study of the particular aspects of the process of fertility decline for selected demographic-social groups (..) On the other hand, a satisfactory interpretation of the factors of fertility decline cannot be made without an accurate reconstruction of a history of mentalities. The gradual acceptance of birth control must be explained in the light of the general evolution of the population's attitudes and mentality. The changes in religious sentiments, and in the acceptance of prevailing morals; the changes in the relations between men and women, and between parents and children; the formation of a political or class consciousness among the rural, industrial, urban or middle classes: these are just a few of the many pieces which make up the jigsaw of the general cultural climate which has produced new attitudes vis-à-vis procreation. The first strategy is the task of the demographer. However, he cannot totally ignore the second, unless he is resigned to accepting a purely deterministic interpretation of demographic facts" (Livi Bacci, 1977).

In order to seek a few answers to these questions, we must extend our usual toolbox beyond the cognitive processes.

2. EXPLORING THE SPHERE OF THE EMOTIONS

It is usually said that cultural models are sufficiently integrated symbolic systems. What is sometimes forgotten is that this integration should not be sought in consistencies of a logical-cognitive type, but in emotional ones. A system is integrated if the same set of affective dispositions is equally satisfied at each point of application of the model itself. It is often also said that values represent a powerful impetus to action: however their effectiveness does not depend on their content but on their affective charge. The exercising of a value is always an affective experience, i.e., one that involves the sphere of feelings. Simmel (1907) writes:

> "Of will it has rightly been said that it is blind, but this statement has also been misunderstood. Its action is not irrational in the sense of reason as a value-laden concept, but it cannot produce any effect if it does not receive some kind of content, which is never included in will itself; indeed, it is none other than one of the psychological forms (such as being, having to, hoping) in which contents live in us, one of the categories through which we comprehend the content of the world in itself, at a purely ideal level, in order that it might take on a practical meaning for us. Just as will does not - on its own - choose any given content, so it does not derive any purpose from pure knowledge of the contents of the world, i.e., from intellectuality (...). Wherever the intellect leads us, we are in a condition of mere dependency, because its path only crosses the objective connections of things (..). Even if we turn to the concept of calculating the means with total clarity, as long as we stop at this, we remain purely theoretical beings, in no way practical. Will is limited to accompanying the series of our reflections like an organ pedal or like the general premise of a field, upon the particular contents and situation of which it does not have the remotest effect, but into which it (and it alone) can introduce life and reality".

The role of the affective components of a cultural system already emerges forcefully in the value models. In order to give it a classification in broad principle, we start by defining the concept of attitude. It is identified by the simultaneous existence of three elements: a subject, its lasting intentional state, and an object or situation that is the specific goal of intent. In an affective structure (Isen, 1984), the third condition of existence of a cognitive structure (orientation of intent towards an object or goal) may or may not be present. For example, depression is an understandable response to certain difficulties in life events (such as the symbolic or real loss of an object of attachment, i.e., bereavement), but in its development it produces a generalized and unfocused disposition of hopelessness, lacking in object and goal (Brown and Harris, 1978). In these pages, we do not intend to deal with emotions, which are reactive and possess an object, but with critical moods and dispositions, which are not structured in function of an object: euphoria and bad temper, depression and aggressiveness, melancholy and apathy, insecurity and impotence, anxiety and avoidance.

In speaking of the influence of the affective states upon choices, we cannot ignore two authoritative references for the theory of rational action. One is that of Leibenstein (1981) who, reflecting on the logic of procreative choices, affirms, "a high proportion of fertility rate is determined by 'non-decision decisions'", and goes on to make the following affirmation:

> "We assume that active decision-making is infrequent. Day-to-day behaviour is on a routinized basis. Passive decisions usually involve routine behaviour. It may be viewed as behaviour 'within a holding pattern' (..). If the event is strong enough and has a significant impact on the assessment of the decision-maker about the outcomes of behaviour, then the impact of the event forces itself on the attention of the decision-making entity. In this way the event may activate the idea that a new, active decision may have to be made (..). We assume that within certain bounds of certain variables the routines in behaviour that occurred in the past simply continue. Only if the changing data go beyond the lower or upper bounds of the inert area does a stimulus arise for active decision-making".

The second reference is the thesis advanced by Etzioni (1998), whereby:

> "The majority of choices people make, including economic ones, are completely or largely based on normative/affective considerations not merely with regard to selection of goals but also of means, and (..) The limited zones in which other, logical-empirical considerations are paramount are themselves defined by normative/affective factors that legitimate and otherwise motivate such decision making"[13].

Other examples of the raising of dispositions to paradigms of social action are to be found into the world of philosophical speculation. Binswanger (1956) matches mannerism as state of would-be existence with Mannerism[14] as philosophical program and artistic current in the late-Renaissance and pre-Baroque period. Georg Simmel vividly depicts the specific traits of the dispositions of the '*blasé*' and the 'cynical' as being consubstantial with a modern urban society centred on money.[15]

Social and economic anthropologists (apparently even further away from psychology, anchored as they are to the elementary structures of social facts) have also turned to cognitive-affective structures. For example, the category of negative reciprocity[16] is, strictly speaking, much more of a predatory disposition than a rational strategy. And Banfield himself attributes dispositional connotations[17] to the concept of familism. In general, the scientific literature (just as much as the narratives of psychological introspection[18]) has elaborated lists of affective structures which influence the action and non-action of individuals and which, with a connotation of crisis, may be assimilated to the concept of orientation of action introduced by Weber in his essay of 1913.

The bearing structure of these cognitive-affective forms may be synthesized in the form of three features, of which the last is particularly important here. The first is the loss of the ability to react, the indifference to stimuli (*akedìa*) as a reaction to the unbearable nature (in quantity or quality) of the stimuli themselves. The second is the inability to choose an order of priorities (Laffi, 1998) and thus the inability to take irreversible decisions. It is the reflection of a paratactic organization of the decision-making process, consisting of the placing alongside each other of possible alternatives, without any identification of interdependencies and priorities (as when a sentence of discourse is constructed by the mere placing together of independent propositions). It is the philosophy of "all and now", without the obligation of choices[19], but with a *post factum* ideological rationalization of these non-decisions[20]. The paratactic degeneration of the decision-making process has something to do with the repulsion of irreversible choices, which scholars of young people's condition in Italy regard as typical of the new generations.[21]

The third basic characteristic of crisis dispositional states is the loss of the nexus between action and goal. Rational choices are choices which are directly goal-oriented. On the contrary, many social or demographic processes break away from the mechanical intention-action model. The choices of crucial passages in the life-course are often choices "not to decide". Elster (1989) defines as "essentially secondary effects" the results of processes lacking in any direct and conscious link between outcome and intention (like crisis dispositions), which require the relaxation of the control of reason in order for them to be fulfilled.[22]

A theoretical category, which well embodies the concept of critical dispositions, is that of security/insecurity of attachment of the child to a figure of reference, which John Bowlby (1969) developed in his two volumes *Attachment and Loss*.[23] Anxious insecurity and avoidant insecurity are two basic mental dispositions which filter and translate individuals' value orientations into behaviour in the affective sphere, and which re-emerge in all peoples critical choices. At the root of recent demographic behaviour in the

Mediterranean regions, there may well be an anxious or avoidant dispositional state, that assumes the role of interceptor between desire and action.[24]

This working hypothesis may be disturbing for the traditional approach to demographic behaviour. In effect, in focussing our analysis of behaviour and attitudes on the underlying drive-dynamics, we alter the methodological habits and conceptual tools of anyone researching the demographic changes underway. Actually, this is an approach that has already been pursued in seminal studies in history and the social sciences. For example, in the chapter in which he synthesizes research on the Authoritarian Personality (Adorno *et al.*, 1973), Sanford wrote: "Behaviour, however consistent it may be, is not the same thing as personality; personality lies behind behaviour and within the individual. The forces of the personality are not responses, but dispositions towards response". The 'forces of the personality' proposed by the Berkeley researchers (thrusts, desires or impulses) correspond to the drives we refer to in these pages. It is notable that for these authors the underlying impulses are of a genesis not assimilable to that of the system of preferences.

Before attempting to operationalize the interpretative categories examined here, we need to make three further points. The first is of an epistemological nature, and underlines how the operational use of dispositions and moods requires us to take a step (backwards?) in the direction of the positivist concept of explanation. Indeed, dispositions and moods operate (and no differently, to tell the truth, from Weber's orientations of action) under a logic similar to causal logic, in a similar way to the dispositional models of explanation laid out by Ryle and taken up again by Hempel (Sparti, 1995),[25] which themselves reproduce the nomological-deductive model at a more sophisticated level. The causes are therefore found outside, in nature, but also inside the human actor, in the form of dispositions to act (or, as in our case, counter-dispositions to act).

Secondly, we would like to distance ourselves from the common-sense definition of crisis dispositions, even slight ones, as affective disorders, which would lead one to believe that there is a qualitative difference between a balanced state of mind, so to speak, and a pathological one. Here, by moods such as depression and anxiety, we mean neutral states of mind rather than pathological forms.[26]

Finally, we must be aware that it is impossible to make a totally clear-cut separation between the cognitive component and the affective one. The scepticism of demographers as regards the potential contribution of psychology to studies on fertility dates back to evaluations of the results of the great American surveys. Fawcett (1970) underlined how neither the Indianapolis studies nor the three rounds of the Princeton survey demonstrated significant correlations between psychological variables and indicators of fertility. There was a mass use of indicators of dispositions and moods in these studies, and none of them (in contrast with indicators of economic status) demonstrated any significant relationship with the demographic data[27]. However, the conclusion of Kiser (1967, quoted in Fawcett, 1970) is relevant to these failures: "the failure to provide associations between demographic behaviour and psychological factors suggests either that the really important psychological features were not identified or that they were not properly measured". Neither value models nor moods work on their own, in the absence of interaction between cognitive and affective factors. Ciompi (1982) is correct in suggesting an interpretation of behaviour according to an "affective logic" (Affektlogik): an integrated system in which cognitive and affective functions form an indivisible unit, simultaneously a logical structure of affects and an affective structure of logic.

3. MEASURING VALUE MODELS AND DISPOSITIONS

The two previously used Italian surveys attempted to translate into empirical indicators certain value models and dispositional states which might be placed in relation to the system of expectations as regards family behaviour (Table 1).[28]

As for the former, three dimensions of the pact of social and generational coexistence were measured:

- A scale, which measures the degree of post-materialism,[29] proposed by Inglehart (1990), based on Maslow's hierarchical theory of needs (1954).
- A scale of measurement of familistic orientation, which measures the priority assigned, in individual strategic decisions, to the collective good or - in contrast - to the 'particular' good of the family to which one belongs:[30] the two poles of a continuum (Ginsborg, 1998), generally labelled with the categories of civicness (or post-familism) and familism.[31]
- A scale - less usual in the analyses of opinion frames - of adherence to the pact of intergenerational solidarity (always understood in an anthropological sense as kinship descent) or conversely of dissolution of this pact.[32]

From behind the three dimensions of post-materialism, post-familism and the inter-generational pact, there emerges the structure of what Caldwell (1982) defines as "family ethics": rules governing who, in the family, may obtain what (and how much) from whom, and who may tell whom what to do.[33]

As post-familism is the overcoming of a mental habit, which is even taken as the stereotype of the 'Italian way of thinking', the overcoming of familistic patterns should accompany the tendency to refuse pacts of solidarity between generations. In contrast, in both surveys the opposite is observed: the rescission of the intergenerational contract is much more often accompanied by the conservation, rather than the abandonment, of familism.

Table 1. Abacus and Friuli surveys: variables used in a component analysis

	ABACUS SURVEY	FRIULI SURVEY
Synchronic background variables	°Age (five-year age groups) °Level of education °Size of city	°Age (three cohorts) °Educational qualifications °Leaps in social status between generations
Diachronic background variables		°Material deprivation in childhood °Lack of domestic harmony in childhood °Affective deprivation in childhood
Intermediate variables: value models	°Rescission of pact between generations °Post-materialism °Post-familism °Political self-definition °Ethnocentricism	°Rescission of pact between generations °Post-materialism °Post-familism
Intermediate variables: dispositional states	°Avoidance °Self-regulation/Anxiety °Typology of forms of insecurity	°Avoidance °Self-regulation/Anxiety °Trustfulness °Tendency to repress problems °Perceived relational malaise °Perceived physical malaise °Need for reversibility °Drive dependency °Solidarity
Output: attitudes	°Proper number of children	°Proper number of children
Output: behaviour	°Number of strong ties	°Actual number of children (compared to standard number for age) °Number of strong ties

In particular, the Abacus survey[34] shows a marked polarization between two sets of propensities, which characterize two blocks of the population that are distinct, both socially and culturally. The first, more educated and modern, is marked by a post-materialist orientation and by dispositional states linked to this (for example, greater tolerance towards the transgression of residual norms[35] recurring in everyday life in the private sphere). The second block, marked by the opposite value models from the former, concerns those sections of the population who are less - or more recently - educated, with little (or no) prospects of social mobility down the generations (which leads to a real condition of relative deprivation).

Apart from these indicators of value models, the Friuli and Abacus surveys attempted to measure certain dispositional states. A measure of the level of trustful openness to intersubjective relations[36] and an indicator of the physical perception of malaise[37] fall into this group. But the most ambitious attempt has been to measure the level of anxious insecurity and avoidant insecurity of an individual, founding categories of John Bowlby's theory of imprinting,[38] with a different procedure from that used by Bowlby's school. In order to do this, we used scales of semantic perception, starting from sets of key words, in order to express the sensation of pleasantness/unpleasantness, which they caused. A sample of 45 words of strong evocative potential, pre-selected for their representativeness and stability of meaning, turned out to be sufficient for the construction of reliable psychocultural clusters (Micheli, 1999).

The matrix of scores assigned to each respondent for each word, according to the pleasantness of the sensation evoked,[39] was submitted to a principal component analysis. Of the first four components, two[40] may be traced back to the states of insecurity in Bowlby's theory: a normative dimension[41], which distinguishes the states of emancipating (but anxious) self-regulation from those of hetero-directed regulation; and a relational dimension, which distinguishes the states of trustfulness from those of avoidant insecurity.[42]

The insertion of the two indicators of insecurity (anxious, avoidant) in the analytical models of expectations and behaviour formation turns out to be fairly useful. For the Abacus survey data (Figures 3 and 4), it emerges that the proper number of children only exceeds replacement level from generation to generation among those with low levels of insecurity, either anxious or avoidant. This finding also holds for each age group taken separately, and it is theoretically reasonable. In demographers' analyses,[43] anxious insecurity is often associated with the tendency to increase the value attributed to the risking of one's own decisions, and consequently to postpone them or contain their choice.

Table 2. Measurement of varimax five principal components.

ABACUS SURVEY Variables- Factors (%)	I 12.85	II 11.56	III 11.52	IV 11.27	V 11.15
Age	+.74	0	+.06	-.08	+.28
Rescission	+.68	-.07	-.05	+.06	-.21
Post-materialism	-.21	+.71	+.14	-.19	+.11
Political self-definition	-.06	*-.69*	+.28	-.21	-.04
Anxious self-regulation	+.22	+.49	+.47	+.11	-.34
Avoidance	+.25	-.05	*-.57*	+.01	-.27
Post-familism	+.12	-.09	+.72	+.05	-.09
Size of city	+.16	+.11	-.12	+.78	+.08
Education	*-.27*	-.11	+.30	+.70	0
Size of net-work	+.29	+.09	-.02	-.08	+.64
Proper number of children	-.20	-.02	+.04	+.18	+.69
Identification of factors	Rescission of generational pact	Anxiety /Post-materialism	Avoidance/ Familism	Education/ Urbanization	Family target/ Size of network

11 dispositional and background variables, contribution to the total variance = 58%

Table 3. Five principal components (12 intermediate variables) and contribution to the total variance, by age cohorts.

FRIULI SURVEY Variables – Factors	Total	1973-75	1959-61	1945-47
° Avoidance ° Rescission of interg. pact	III (12)	I (12)	I (13)	III (12)
° Anxious self-regulation ° Need for reversibility	II (12)	II (12)	V (10)	II (13)
° Relational malaise ° (Decline in) trustfulness	I (13)	III (12)	III (12)	I (15)
° Post-materialism	IV (10)	V (10)	II (13)	IV (11)
° Post-familism ° (No) tendency to repress	V (9)	IV (10)	IV (12)	V (9)

Figure 3. Abacus survey: proper number of children, by level of anxiousness and gender

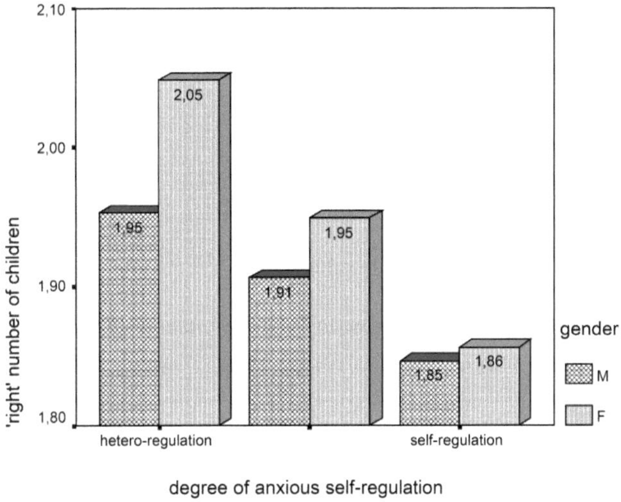

Figure 4. Abacus survey: proper number of children, by level of avoidant insecurity and gender

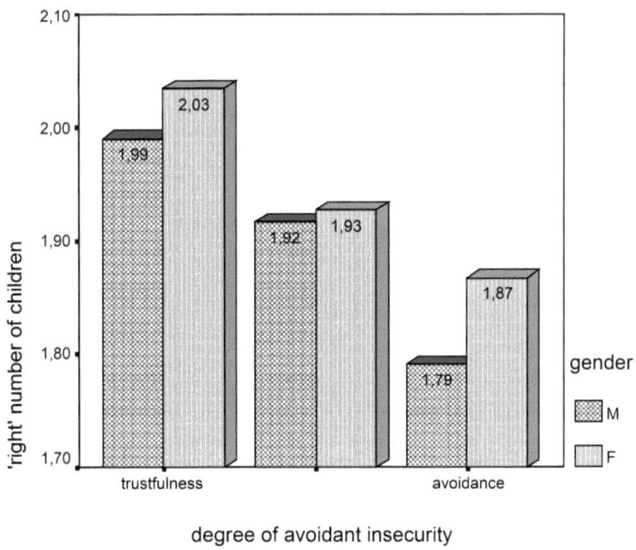

Table 4. Multiple regression for proper number of children.

ABACUS SURVEY	M+F	M	F
Size of network	+ *	+	+ *
Age	+	-	+
Anxious insecurity	- ***	-	- ***
Avoidance	- ***	- *	- **
Size of city	+	-	+
Education	+	-	+
Significance (F)	***		***
N	(1085)	(363)	(721)

* P-value < .05 ** P-value < .01 *** P-value < .001

Table 5. Multiple regression for actual number of children.

ABACUS SURVEY	M+F	M	F
Size of network	+	+	+
Age	+ *	+	+
Anxious insecurity	- ***	- *	- **
Avoidance	- **	- **	-
Size of city	+	+	+
Education	+	+ *	-
Significance (F)	***	***	*
N	(371)	(124)	(247)

* P-value < .05 ** P-value < .01 *** P-value < .001

Table 6. Multiple regression for proper and actual number of children, by age.

FRIULI SURVEY	Proper no. of children		Actual no. of children	
	38 years	52 years	38 years	52 years
Education	+	-	-	- *
Relational malaise	-	+	- **	-
Rescission of generational pact	- **	-	- **	-
Anxious insecurity	- *	-	-	- *
Post-materialism	-	-	-	+
Post-familism	+	-	-	+
Significance (F)	*		***	**
N	(160)	(162)	(165)	(162)

* P-value < .05 ** P-value < .01 *** P-value < .001

Tables 4, 5 and 6, with regression models applied to the data of the two surveys,[44] are in line with this interpretation. In the Abacus data, intention as regards the proper number of children (Table 4) is influenced much less by the background variables, and much more by the dispositions of anxiety and avoidance, and this model turns out to be highly significant for the women interviewed. If we insert the actual number of children into the explanatory variables, this variable absorbs the model's entire contribution to the total variance: actual number of children and proper number of children proceed in parallel, partly because of a selection effect and partly because of a mechanism of *post factum* rationalization, such as those suggested in figure 1b. But if we apply the same regression model to the actual number of children, the overall model is not significantly modified (Table 5): the dispositional factors are still dominant compared to the structural ones and turn out, surprisingly, to be even more significant than the factor of age.

A similar model applied to the data of the Friuli survey (where the indicator of avoidance is effectively substituted with the variable 'rescission of the generational pact') exhibits a lower degree of statistical significance. But even in this case, it was dispositional states (rather than explanatory variables relative to value models), which exhibited a statistically significant influence on actual behaviour.

4. WHAT LEADS TO A CHANGE IN DISPOSITIONAL STATES?

In the previous pages, we cultivated the doubt that in order to understand a change in behaviour, social practices or attitudes, it is not sufficient to seek - as it would be according to the more usual dictates of cognitive psychology - a mere change in objective parameters or in their knowledge. We need equally[45] to ask ourselves if and how a change in value models (or in dispositional states acting as a gyroscope)[46] might lie behind the change in social practices. But what critical events, individual or collective, might themselves explain this type of changes? And through what procedures?

In order to attempt to answer, let us take two steps backwards. First, let us seek in psychological literature a sufficiently comprehensive model of compensatory mechanism for the coping with critical events. Let us then classify possible critical events according to a (rough) typology, and see whether different types of critical events correspond to different types of compensatory mechanism. Social psychologists have identified a powerful mechanism for explaining changes in behaviour and preferences. This includes anticipatory self-reinforcement (Heckausen, 1999)[47] and cognitive dissonance, a theory of Festinger's (1957) synthesized by Elster (1999):

> "Whenever a tension among the elements of a person's mental set generates psychic discomfort, 'something has to give'. A pressure is set up to adjust one or more of the elements in the

set so as to reduce the tension or dissonance and restore harmony or consonance".[48]

Although the mechanism of cognitive dissonance is intrinsically unconscious, it has generally been applied to the alteration of behaviour and preferences, hence well within the sphere of cognitive constructs. But its sphere of application is broader: Heckausen (1999) speaks extensively of action on the mental representations.[49] Not just cognitive constructs, then, but also normative systems.[50] With a further crucial extension, the theory of dissonance is also extended in these pages to fuzzy and non goal-oriented dispositions.[51]

Ciompi (1982) has highlighted a distinction between dissonances by contradiction and by paradox,[52] that could be important in predicting the type of adaptive reaction of the individual: indeed, a contradiction induces an adaptive reaction of restructuring of the cognitive system, while a paradox channels the response more drastically towards the destructuring of the affective system, with an effect which is possibly more extreme.[53] The quality of critical events and situations capable of producing dissonances (in one or the other form) is therefore crucial for predicting the type of reaction. Two criteria of classification of events and situations emerge as being particularly useful.

The first rule starts from the individual/collective dichotomy. Festinger's theory generally gave priority to the psychological-social dimension, in which individual or collective behaviour or attitudes are influenced by individual contingencies, generating dissonance. But it also turns out to be useful in interpreting the effect on collective behaviour of collective events or contingencies. According to an ethological interpretative scheme (Eibl Eibesfeld, 1983), any historical discontinuities in the rules of social control may be the result of the worsening of social aggressiveness. In a recent interview, Luis Sepulveda noted impressionistically (but correctly) how, upon returning to Chile after Pinochet, he had found a change in the climate of human and trusts relationships, with the foundering of the trust capital "typical of his people", and its substitution by a tendency towards the monetization of human relationships. Similarly, Simmel (1903) represents the formation of an affectively neutral metropolitan mood as an effect of a collective condition of dissonance, due to a sort of "environmental desertification".[54]

The second rule starts from the synchronic/diachronic dichotomy, and it has to do with the lapse of time between the contingency that generates dissonance and the formation of a change in dispositional states. Alongside mechanisms of strict synchronism between cause and effect, there may indeed exist at least two important types of asynchrony: a kind of synchronism in two stages, within the same life course, and a diachronic mechanism which straddles two generations.

Examples of two-stage mechanisms recur in the psychological literature with a phenomenological approach. The two following passages throw light on the quite distinct mechanics working at two stages. The first consists of a reactive attitude (e.g., depression as numbing of feelings) towards an unbearable event, like the loss of a figure of reference (Brown and Harris, 1978), or towards a situation that is unbearably contradictory, such as the lack of a home, both a physical and symbolic place of reference (Binswanger, 1956). The second stage consists of the generalization and destructuring of the reactive state in a dispositional state lacking in object and goal (hopelessness, mannerism).

> "Like Bowlby, we believe hopelessness is the key factor in the genesis of clinical depression, and loss is probably the most likely cause of profound hopelessness. The immediate response to loss of an important source of positive value is likely to be a sense of hopelessness, accompanied by a gamut of feelings, ranging from distress, depression and shame to anger. Feelings of hopelessness will not always be restricted to the provoking incident – large or small. It may lead to thoughts about the hopelessness of one's life in general. It is such a generalization of hopelessness that we believe forms the central core of a depressive disorder" (Brown and Harris, 1978).

> "The university student Jurg Zund is lively, sensitive and impulsive as a boy, but he suffers from states of anxiousness and abnormal bodily sensations. Soon he moves in three 'worlds' that are different or downright contradictory. He fails to put down roots in any of these 'worlds', given that he constantly sees each of them in the mirror of the other two (..). The existence of Jurg Zund is threatened right from the start by a schism, by a triplicity of directions" (Binswanger, 1956).

It is not only analyses of a clinical type that might benefit from this mechanism of interpretation. Let us attempt to use it to explain the differences between cohorts in the intentions and dispositions of respondents in the Friuli survey. Here, in the formation of the substrate of current behaviour, an important role is played by the great social transformation (measured through improvements in educational qualifications from mothers to daughters), which remodelled the region throughout the cohorts interviewed. In just a few decades, the region passed from a scenario of substantial immobility to one of explosive upward mobility, in which a stationary life history takes on a connotation of social decline. And the part of the population, which remains left out of the new standards of social status, experiences a growing malaise.

Table 7. Regression relating anxious insecurity and rescission of the intergenerational pact by various background variables[55], by respondent's age group.

FRIULI SURVEY	Anxious insecurity		Rescission of the pact between generations	
	38 years	52 years	38 years	52 years
Level of education	+ ***	+ ***	-	+
Generational social mobility	-	- ***	-	-
Material deprivation	-	+	-	-
Lack of domestic harmony	+	-	+ *	-
Affective deprivation	-	+	+	+ *
Significance (F)	*	***	**	
N	(164)	(163)	(162)	(164)

* P-value < .05 ** P-value < .01 *** P-value < .001

It is therefore reasonable that (Table 7) anxious insecurity should appear especially in those sections which are educated but do not benefit from the collective post-war process of upward mobility: sections basically characterized by a certain status inconsistency. And nor is it surprising that those minorities with stationary life histories in contexts of collective upward mobility, marked by the low quality of affective life experienced in their childhoods, should display syndromes of avoidance and rejection of the intergenerational solidarity pact (Micheli, 1999b).

Indeed, the most frequent symptoms of affective deprivation,[56] potentially generating a disturbed imprinting, are to be seen precisely among those women born and raised in the 1960s in a family context of a locked social mobility. And there appears to be a correlation between these symptoms and frames of orientation towards action, which are marked by six interwoven aspects, all more or less present in the critical generations explored: a) a sharp reduction of the network of strong ties; b) a marked fall in the level of trustfulness; c) a strong impulse towards avoidant insecurity; d) the physical perception of relational malaise; e) rescission of the solidarity pact between parents and children; f) a drastic fall in the interest declared in the experience of parenthood.[57]

Let us now move on to the diachronic mechanisms of dissonance. In this case, the two steps of the mechanism (cognitive dissonance and change of the mood) do not follow upon one another within the life history of the same individual, but bridge two successive generations, as the baton passes in a relay from parents to children. Bowlby's categories of anxious and avoidant insecurity provide a highly illustrative example. Indeed, reflecting on the second demographic transition in Italy, we may ask ourselves whether the ideational change really took place in concomitance with a dissonance effect, or rather some time after. It is no coincidence that the cohorts of women born after 1964, who set off the process of decline in reproductive behaviour, are in fact (on average) the daughters of couples formed during the hardship of the post-war years. Roussel (1992) was the first to stress how 1965 coincides with the entry into adulthood of those generations born at the end of or just after the second world war. In his words,

> "It is probable that these young people were exposed to a model of socialization different from that of their older siblings. The families, which brought them up, had new attitudes towards institutions and the symbolic systems, which legitimise them, and these naturally influenced the children. So the real origin of the changes would appear to lie in 1940 rather than in 1965. What is certain is that the generation-effect seems undeniable".

We may synthesize the problem first highlighted by Roussel in the following way. The generation that goes through the stage of family formation in the years 1945-50 experiences, in a critical phase of its own life cycle, a context of great *anomie*; as a result, it alters its own philosophy of life, in accordance with the theory of cognitive dissonance. But the change of reproductive models as a result of a climate of *anomie* should either only concern the adult cohorts in that historical contingency, or reverberate, by a mechanism of anticipatory socialization, on all the younger cohorts also present (the younger siblings). Change should spread, so to speak, through contiguous cohorts like an epidemic. But the cohorts entering into adult life in the 1950s (Roussel, 1992) adopt basically traditional family models. The change does not affect the already semi-adult cohorts in the post-war period (protagonists over the next few years of the baby boom), but only those who were in their earliest childhood at the time, and enter adult life in the seventies.

For a change in the logic of action to take root only from a certain cohort onwards requires a cohort-specific transmission of the experience. Bowlby's paradigm of attachment (1969) offers a satisfactory interpretation: the selective transmission is the result of a disturbed activation of the imprinting process.[58] More in detail, the feeling of attachment in the early years of life, experienced in an anomic context by the generation born in the immediate post-war period, itself produced (precisely and only in this cohort) either a change in (anxious insecurity) or a partial de-activation of (avoidant insecurity) both the instinct of attachment and its interface, the instinct of care. And an anxious or avoidant assumption of the instinct of care may explain the recent changes in the transition to the adulthood.

The growing tendency to remain in the niche of a pre-adult state therefore finds an additional (though not exclusive) explanation in a sort of locking of dispositional states: something like the disengaging of the clutch, which makes it impossible to make a vehicle move forward, however desperately we put our foot down on the accelerator. The tendency to put off exiting from childhood indefinitely is therefore explained not (or not only) by the many non-pull factors present in the literature (limited attractiveness of the adult world, excessive costs of transition), but also (or even rather) by non-push factors.

The suggestion advanced in these pages of linking the demographic transformations underway in Italy to the causal chain "disturbed imprinting → anxious or de-activated instinct of care" may appear out of the mainstream of demographic debate, but it is by no means isolated within the broader sphere of historical and social studies. Stone (1983) and Trumbach (1982) apply it to the systems of trust-based relationships in the *ancien régime*. And some vicious circles between poverty and overpopulation typical of the excluded peripheries of the large urban centres of Latin America, or the Black slums, may be interpreted in terms of the de-activation of responsible parenthood among young women and men who have undergone the experience of abandonment and unfulfilled imprinting in early childhood.

5. SUMMARY AND THREE POSSIBLE LINES OF FURTHER ENQUIRY

Once the category of familism had been split up into two distinct philosophies, one competitive in the market of families, the other solidarity-based within the stem-family, these pages first of all highlighted an important change in the Latin and Mediterranean model. This has developed from the traditional amoral familism, stable over time, to a sort of amoral individualism inscribed within a demographically collapsed society, itself also a robust and stable model. But what is it that explains the dismantling and renewal of the system of practices and social norms, which has taken place in recent decades? We looked for an alternative interpretation to the usual socio-economic one, exploring the signs of a change in the collective framework of symbolic and emotional references. We sought to enlarge our toolbox beyond cognitive processes, exploring the role of the affective components in a crisis, characterized by the loss of reactive capacity, paratactical organization of the decision-making process and dissolution of the link between action and goal.

The operationalization of the analytical categories of value models and moods, starting from some Italian surveys, provided us with modest results, in the absence of any appropriate longitudinal research projects, but this helped us to formulate alternative, theory-laden interpretations of the demographic processes underway in Italy. We therefore strove to extend our knowledge of the possible constituent devices of the explanatory models. In particular, we sought to highlight the usefulness of studying both the two-stage mechanisms of change of dispositional states[59] and the climates and situations in which human agency is incubated, and also the mechanisms governing the degree of transmission of symbolic heritage from generation to generation. However, this approach opens up theoretical and operational problems. I shall limit myself, in conclusion, to indicating three orders of problems.

First of all, dispositional states can act as intermediate variables between incubating climates and transformation of social practices, and their change can be decisive in the decision-making process. However, we sometimes observe a change in value models and states of mind in the presence of certain individual or collective contingencies, sometimes - in the presence of the same contingencies - there is no change, or the changes are in the opposite direction. What is the reason for this? The reagent might lie in the very anthropological structure of primary relations, which is itself the end product of practices consolidated over time. The rescission of the pact between generations would take place in those very regions where previously there existed a fabric of close-knit networks of ties; the loosening of these ties would be the *vulnus*, which might produce an equal, and opposite reaction (see Micheli, first contribution in this book).

The second line of further enquiry departs from the observation that the processes of mass education have not brought about a homogenisation (either upward or downward) of the population. In contrast, the country currently seems to be divided, according to acculturation and relational capability, into two Italy's, both undergoing change. The more educated and integrated one is sailing under the flag of modernization, post-familism and post-materialism. The less educated and less integrated one (scattered throughout the peninsula and therefore also across the regions of the north), is subject to a different kind of modernization, which combines persistent familistic values with manifestations of disruption in both interpersonal relationships and in intergenerational ones (ethnocentrism, intolerance, rescission of the generational pact). The two forms of insecurity are also differently distributed in the two Italy's. In all age groups, the more educated classes are characterized by a disposition of anxious self-regulation, while the less educated sections are characterized by a state of avoidant insecurity.

Figure 5. How different dispositional states and logical-affective dynamics, working in different social niches, can produce apparently similar demographic behaviour

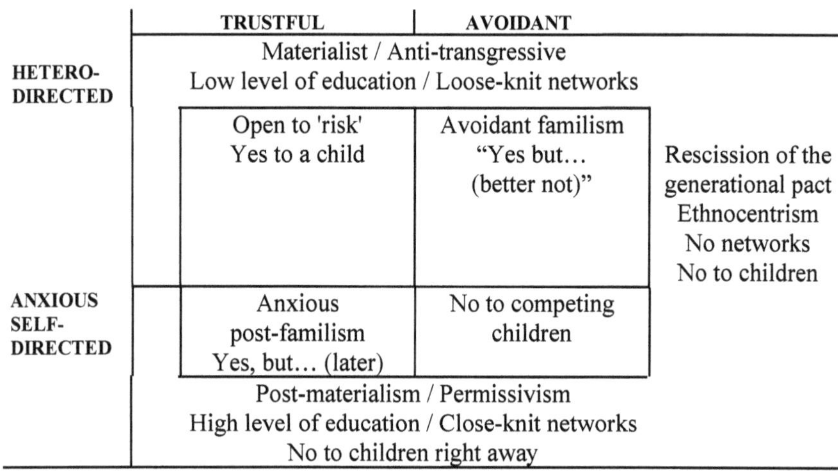

Two distinct populations are taking form, with two quite distinct models of norms, values and dispositions (Figure 5) – Micheli, 1999a. A low degree of education, poor relational capability, the persistence of materialistic priorities and familist values and a tendency to break the generational pact go hand in hand with a general state of avoidant insecurity and a widespread desire to deny that procreation is the only way to self-fulfilment. Vice versa, a high level of education, a close-knit network of ties and a preference for post-materialist and post-familist priorities are the ingredients of a self-directed modernity, steeped in anxious insecurity. This second recipe corresponds to an attitude, which is generally more favourable towards procreative choices: the proper number of children is greater for post-familists and post-materialists in almost all age groups. But in this case, irreversible choices of transition are subject to anxious overvaluation, which will lead to their indefinite postponement, right up to the limit of total renunciation.

It would therefore appear fairly legitimate to conclude that the current demographic trends in Italy (prolonged containment of births, putting off of all the passages of the life cycle) are not the result of a single - however complex - process of social transformation, so much as the identical manifestation of two quite distinct processes taking place in two Italy's which are sociologically distinct.

The third and last line of enquiry starts from the downright obvious observation that crisis in the Mediterranean models of transition to the adulthood is not an isolated process: it can not be differentiated from the great fault-lines characterizing other crisis and transitional behaviour in central-Mediterranean Europe at this current turn of century. From behind reasonable localist aspirations are emerging impulses of ethnocentrism, which we hoped were long buried, embedded into forms of collective intolerance. New experiences of searching for extremes and new experiments in the 'disconfirmation' of social life and life itself are making headway. There is an increase in the prevalence and a change in the significance of juvenile suicide, while new anxiety syndromes spread rapidly, generally in the form of eating disorders.

There is certainly a bridge between demographic behaviour and crisis behaviour. If we were able to provide a 'macroscopic' interpretation of the processes underway, exploring the role of logical-affective components both in crisis behaviour and in the normal social practices of demographic reproduction, we would have more elements at hand to better understand the tidal wave of modernization in which we are immersed.

Lesthaeghe (1986) described the second demographic transition as being an essentially liberating process.[60] In these pages, highlighting the role of critical dispositional states in the transformations underway, we give what is actually a less optimistic vision of the general process of modernization. This does not mean to reject the emancipating and secularising side of modernization, enfolding us in a Gothic embrace of the irrational. But the balancing of light and shade, in the reading of historical processes, cannot but lead to a better appreciation of their depth of perspective.

Giuseppe A. Micheli, Institute of Population and Geographical Studies, Catholic University of Sacred Heart, Milan.

NOTES

[1] The Friuli-Venezia Giulia survey, conducted in 1998, explored the norms perceived by women for the events of family formation, using a sample of 511 women belonging to three 3-year cohorts (1945-47, 1959-61 and 1973-75). Details regarding the design of the survey may be found in Billari and Micheli (1999). Friuli-Venezia Giulia is one of the regions with the lowest rates of fertility, traditionally settled by the stem family, as reflected in the late age of exit from the family of origin and the existence of wide network of family businesses. The paternal home represents the main group of normative reference for the young person (Billari and Micheli, 2001).

[2] See Micheli, first contribution, note 34.

[3] A panel designed to explore the social interaction of young adults in the Netherlands (Liebfroer, 1998), shows how "the young adults intending to experience family life even in the relatively near future generally experience this event earlier than young adults who have no short-term intentions" and how "this correspondence between intention and behaviour is stronger for leaving home and childbearing than for union formation and marriage". However, intentions and behaviour display a stronger correspondence in the case of imminent events of family formation. The connection is weaker for events that are further away in time and in sequence of events. So here we have a relatively limited ability to predict.

[4] Casacchia and Dalla Zuanna (1998) show that there is no precise trend per cohort at a national level for the link between parent and child fertility. In reality, several general social surveys conducted in Italy in the 1990s on a regional scale (Emilia, Lombardia, the Veneto) regularly demonstrated the strong viscosity of children's family models compared to those of their parents. These confirm the frequently noted and scientifically approved mechanism of anticipatory socialization (Balbo, May and Micheli, 1990; IreR, 1991; Mauri and Micheli, 1992).

[5] This interpretation is supported by the observation that the 1970s saw the concomitance of two processes which were both possible consequences of that perception of incommensurability of cost: a decrease in the number of costly choices (fewer marriages, fewer children) and the gradual dismantling of normative systems based on the irreversibility of family agencies.

[6] The category of attitude develops in the shadow of three different traditions (*fin de siècle* experimental psychology, psychodynamics and the Chicago school). In 1935 Allport adopted it as in indispensable tool, both for social psychology and for the psychology of the personality, even though it was lacking in any clear scientific definition (Rokeach, 1973). With the great empirical studies of the Frankfurt school in the 1940s, the use of social attitudes for the theory of the personality shifts the conceptualisation towards the category of drive, to which we shall return.

[7] An initial operational critique was of the tendency towards superficial identification as attitude of everything that the subject declares as his attitude.

[8] In these pages, the term norm or normative system is used to indicate a broad range of meanings. It embraces the various elements of a culture, since they all effectively manifest themselves as normative information, that induce people, through channels

of varying degrees of formalization, to adopt the ways of acting, thinking and feeling practiced in a community. Complex constructs such as that of values (collective standards which furnish criteria of judgment on ways of acting, thinking or feeling) and models or 'value patterns' (concrete and complex constructs of simple values) fall under this broad-ranging concept. Such models include, for example, the new childrearing ideology referred to by Gullestad (1997) (the shift "from a rhetorical emphasis on 'being of use' to an emphasis on 'being oneself' in the upbringing of children", within which "there is an opposition between 'being oneself' and 'being obedient'"), and Banfield's concept (1958) of familism as a syndrome("some readers may feel that amoral familism, or something much akin to it, exists in every society, the American no less than the Southern Italian. Our answer to this is that amoral familism is a pattern or syndrome; a society exhibiting some of the constituent elements of the syndrome is decisively different from one exhibiting all of them together"), and the typology of Atlantic, Latin and Mediterranean family models used in Micheli (first contribution). In this latter range of meanings, the dialectic between norms and social practices (habits and conventions of daily life which, while not incorporated in formal rules, nonetheless generate pressure to conform, for the actor, and expectations of such conformity from his counterparts) is a crucial aspect of change. For example, to what extent is the proper age for certain passages to adulthood an established social norm and to what extent is it, in contrast, a social practice undergoing slow transformation, capable of pressure as it changes? I believe it is important to throw light on these mechanisms, constructing research plans which are not only quantitative.

[9] For the time being, let us cite, by way of example of this mechanism, the paralysing tendency to avoid the irreversible in effective choices. We shall fill in the details as we go along.

[10] The endocrine systems make use of chemical messengers, hormones, which are capable of acting at a distance. In the nervous system, chemical messengers, the neuro-mediators, ensure synaptic communication between the nervous cells. The target (receptor) is a molecule situated on the cellular membrane of the hormones. The affinity of a hormone for its receptor is the force that brings them together and unites them. While it is selective, the affinity between a receptor and its hormone may not be exclusive. Other ("analogous") substances may be recognized by the receptor, and they unite with it, taking the place of the hormone. If this union gives rise to an effect similar to the one produced by the hormone, the substance is known as an "agonist"; if, on the other hand, the analogous substance takes the place of the hormone without producing this effect, it is called an "antagonist".

[11] "One should be very careful, at any rate, before attributing to actions black-box causes that are neither rational nor even meaningful to the actor.." (Boudon, 1998). Boudon does not exclude the Weberian typology, which admits of rational social actions not only with respect to a goal, but also with respect to a value, to an affective dispositional state or to acquired habits. However, in sticking to rigidly cognitivist premises (Boudon, 1995) and in the name of the principle of methodological individualism, he drowns them all in the swamp of "irrational causes".

[12] As for the explanation of the change of the framework, there is nothing in the way of our attempting to seek its structural coordinates: this, however, will require access to information which goes back in time, to the formative climates of mutagenesis of the frame. Within the extremely modest limits posed by a survey with retrospective questions, this is what I shall attempt to sketch out in section 4.

[13] Etzioni affirms that "normative/affective factors influence the selection of means by many ways", but he underlines three of these in particular (*ibid.*) "excluding the role of logical-empirical considerations in many areas; infusing the deliberations in such a way that logical-empirical considerations play a relatively minor or secondary role to normative/affective factors; (fencing) the rational decision making into specific normative/affective indifference zones"

[14] Characteristic of a "state of would-be existence" is the loss of a centre, or even of a whole way of life, and subordination to whatever model is currently on offer from others, the adoption of the *allures* whatever "world" presents itself.

[15] "The essence of the *blasé* consists in the blunting of the sensibility towards the differences between things. The meaning and value of differences, and - alongside this - the meaning and value of the things themselves, are perceived as being irrelevant. Everything appears in a uniform colour: grey, opaque and incapable of giving rise to preferences" (Simmel, 1903). "The crucial distinction is not so much the undervaluing of things, as indifference towards their specific differences (...). In the mental disposition of the cynic, experience of the great quantity of things which can be obtained with money, and the consequent reasoning which establishes that everything and everyone has its price, gives rise to a positive sense of joy; for those inclined towards a *blasé* attitude, on the other hand, the same image of reality is stripped of all attraction" (Simmel, 1907).

[16] Negative reciprocity is a form of exchange, which features as "the attempt to get away with obtaining something in exchange for nothing". The concept was coined by Sahlins (1972) as an extension of the thought of Polanyi (1944).

[17] "When the peasant speaks of *la miseria*, he refers first to his hard physical labour, to his patched rags, and to the bread that is often all he has to eat. Cruel as it is, however, his poverty does not entirely account for his chronic melancholy (..). In part, the peasant's melancholy is caused by worry. Having no savings, he must always dread what is likely to happen (..). But neither his present hunger nor his anticipation of worse to come fully accounts for the peasant's deep dissatisfaction (..). What makes the difference between a low level of living and *la miseria* comes from culture. Unlike the primitive, the peasant feels himself part of a larger society which he is 'in' but not altogether 'of'" (Banfield, 1958).

[18] For example, Lucretius speaks of a melancholy or restlessness which men are unable to name. Kierkegaard defines as *ennui* a feeling without feeling, affectivity without affection, the impossibility of deciding. Handke refers to "unhappiness without desires".

[19] "We are perhaps too used to regarding fertility as the fruit of decisions aimed at maximizing individual usefulness or, at the most, that of the couple, almost as though the interest of the (future?) child ran against that of the parents. In contrast, one in-

terpretation of very low fertility in Italy perhaps lies in the attempt to maximize the functions of usefulness of each of the two partners at the same time; there is a simultaneous desire for the unrivalled supremacy of couple, wife, husband and child" (Casacchia and Dalla Zuanna, 1988).

[20] "With the loosening of traditional factors, we have had a multiplication of elements to consider in the decision-making process. Any reproductive choice finds justifications, which are already nice and ready at a collective level (...). There appears to exist a normative ambivalence and a coexistence of value systems which are different and equally valid and recognized" (Palomba, 1991).

[21] "The tension with the reversibility of choices, the main model of reference of an uncertain and contradictory society, seems also to accompany the young person in the important decisions which should affect his own future. It is probable that the putting off of certain crucial choices, such as marriage or procreation, has its origin in the fact that they seem to be irreversible events" (Buzzi, 1997).

[22] Sleeping, for someone who suffers from insomnia, or being spontaneous, for someone who never is, are examples of states of being (or behaviour) which cannot be reached through actions which are directly goal-oriented, i.e., they are the essentially secondary effects of actions which are not aimed directly at this goal. The direct imposition upon oneself or others of the achievement of such goals is the equivalent of pronouncing paradoxical imperatives, placing the subject in front of alarming scenarios of 'double bind'. In all cases of this kind, the object of desire is a state of partial relaxation of control by our senses, our super ego, and our deepest system of meanings. Having a/another child is also an essentially secondary effect. It is possible to decide not to have another child, but it is rarely (in contexts of rationality which is in a certain sense anomalous) decided, or planned, to have one (except when difficulties have been encountered in having one: the reverse essentially secondary effect). Like sleep, parenthood can be achieved along with a loosening of restrictive grip of the control of rationality.

[23] Brown and Harris (1978) suggest that Bowlby's security of attachment refers to the same processes of childhood for which Benedek (1938) speaks of "trust relationship" and Erickson (1950) of "basic trust".

[24] This is certainly not to deny the existence of a broad range of structural factors - economic, sociological and anthropological - which are decisive in inducing the most recent generations to make self-delimiting choices in the reproductive field, interpreted as rational choices. Here, rather, we mean to insinuate the worm of doubt into the self-sufficiency of some of the main psycho-economic interpretations.

[25] The scheme is the following: first one describes the situation in which the actor is situated, designating an initial cause or condition. Then one ascribes to the actor a general disposition towards a certain type of behaviour. By disposition we mean a characteristic property, which predisposes the actor to behave in certain ways, in a similar fashion to a sugar-lump's disposition to dissolve.

[26] The medicalisation of unhappiness in terms of depression has been defined one of the great disasters of the 20th century (Oakley, 1984). Indeed, it reduces the disease to its organic component, with little attention to the processes of social interaction

involved. The biological component of mood should not be understood simplistically as cause of the mood itself (there is no proof of unidirectional causation between alterations in the biochemical equilibrium - a high level of seratonin or cathecolamin in the body - and states of depression), but if anything as a possible concomitant factor, or also an effect. It is also legitimate to posit the reverse causal direction, considering many pathological categories as states of mind associated with "definitions of situations".

[27] The personality traits used in the Indianapolis study included perception of the interference of the child with one's own independence, fear of pregnancies, the tendency to plan ahead, interest in religion, adherence to traditions, conformity to group models and the sense of personal inadequacy. In the final report of the Princeton study (Westoff *et al.* 1961), the main personality variables are moods: generalized anxiety, need for parenthood, compulsiveness, tolerance of ambiguity, cooperativeness and need-achievement. In the first volume of the report, there are very low correlations between these indicators and the two dependent variables (desired number of children and efficacy of family planning), repeating the failure of Indianapolis. In the second volume, the measures of personality are not even recorded, because "no significant associations were found from any analytical point of view".

[28] Only a longitudinal study makes it possible to verify the ability of models and dispositions as determinants of actual decision-making processes, and not just of their expectations. In the Friuli and Abacus surveys, it was only possible to map connections with already complete behaviour (excess number of children compared to specific standard by age), with social practices underway (size of network of strong ties) and with expectations (proper number of children). This does not allow us to separate the effect of selection from that of socialization (Lesthaeghe and Moors, 1995). However, the mechanism of *post factum* rationalization is a process much more complex to define if applied not to declared expectations, but to semi-projective sets of items, concerning fuzzy, non goal-oriented dispositional states.

[29] This places individuals on a scale of priority of needs, which ranges from the level of primary needs, or survival, through those material needs of stable social organization, to the so-called post-material needs of self-expression and self-fulfilment. The Friuli and Abacus surveys took up the reduced version of the set of items used by Inglehart, in order to construct a fairly broad index. The reduced set used by Inglehart is the following: "There has been much talk in recent years of what the goals of this country should be. Which of the following questions do you regard as being first, second and third in order of importance? a) Maintaining the country's law and order; b) Combating price rises; c) Giving people more decision-making power; d) Protecting freedom of speech". As is fairly obvious, priorities a) and b) are of a material order, while c) and d) are post-material. The order indicated by the respondent also features a greater propensity towards one or other of the two levels.

[30] Let us consider as a proxy of familism a composite indicator based on two items: a) "we are responsible first towards family and children and then towards the community"; and b) "love towards family and children does not justify acting against the interests of the community". Let us define as familistic attitude the priority given to

the good of the family compared to that of the community (in accordance with item a) and as post-familistic the inversion of that priority (in accordance with item b)). In the Friuli survey, only item b) was used.

[31] A curious and little-known connection between particularistic logic and dissolution of the inter-generational pact in demographic behaviour appeared as early as five centuries ago, with reference to the case of Italy, in Guicciardini (1970): "One cannot blame the appetite for having children, because it is natural; but I am right in saying that not having any is a kind of happiness, because even those who have them hearty and hale have doubtless much more displeasure from them than consolation".

[32] The degree of adherence-rejection of the intergenerational pact has been measured by two items (opposite in meaning) which make reference to the obligations of children towards parents and vice versa: a) "It is right for children to make sacrifices and have elderly parents living with them"; b) "Parents have a right to their own lives, without making sacrifices for their children".

[33] In addition to these three dimensions of the 'social contract', other indicators have also been taken into account in these pages. One example is an index of ethnocentrism, defined as the degree of intolerance towards excluded ways of living: e.g., drug rehabilitation centres, hostels for the mentally ill, special workshops for the disabled or travellers' communities.

[34] For data and analyses from the two surveys, see Micheli (1999a, 1999b) and Billari and Micheli (1999).

[35] By residual norms we mean those not regarded as crimes, which form part of a fuzzy sphere, not structured by social censure.

[36] The rate of trustfulness is measured by combining three scales of agreement/disagreement with three statements concerning the relationship of trust both with others and in front of unpredictable future events: a) Most people are worthy of trust; b) It is natural that people should try to take advantage of my good faith, given half a chance c) When I think of my future I see it as being full of risks and unknowns.

[37] The somatised perception of malaise was measured on the basis of three items extracted from the Nottingham Health Profile scale concerning self-perception of the state of health. The conditions or dispositions for which it was asked to indicate if they corresponded to one's own personal state in the recent past are "a) I always feel tired at the moment; b) Doing anything is an effort; c) I get tired easily".

[38] We shall return in Section 4 to the subject of the usefulness of Bowlby's clinical categories in understanding the role of models and moods in the formation of demographic behaviour.

[39] The coordinates on the semantic space may be a good approximation of the affective dimension. Also in Parsons, one of the five pattern variables consists of the axis "immediate gratification (affectivity) - deferred gratification (affective neutrality)". The desirability of tracing the map of dispositional states, gauging the position of semantic perception on the pleasant-unpleasant continuum is corroborated by the findings of Davitz (1969). He studied the communalities among the verbal description of emotional states, through a factor analysis conducted on 215 terms used to

indicate three emotional experiences, and identified four main bipolar dimensions: of these, the two most stable were the level of activation (arousal) and the hedonic tone (pleasantness/unpleasantness) of the emotional state. A similar result already existed in Nowlis and Nowlis (1956), while for Russell (1980), the map of the emotions is a space inscribed in a circle that maintains pleasantness and activation as its two main axes.

[40] The other two latent dimensions identified (marginal in the subsequent connections investigated here) are tied to the dispositional categories of drive dependency (contribution to the total variance = 11.0%), which distinguishes behaviour dominated by drives of order from other behaviour dominated by existential drives, and solidarity (contribution to the total variance = 9.4%), which separates life-philosophies guided by utilitarian logic and competitiveness from others guided by solidarity and an instinct for care).

[41] The first latent axis as per contribution to the total variance (15.7%), it is characterized on the positive axis by clusters of words centred on dispositional states of elusion (evasion, irony, anti-conformism), of reversibility (flexibility, lightness, change), of rationality (logic, reason, construct) and on the negative axis by clusters of rituality (faith, sacred, ceremony), family (marriage, home, faithfulness) and conservatism (hero, law, moral).

[42] The fourth latent axis according to the contribution to the total variance (8.5%), it is characterized on the positive axis by the clusters of avoidance (criticize, rigid, diffidence) and utilitarianism (property, money, wealth) and on the negative axis by the clusters of transcendence (infinite, soul, eternal) and rituality.

[43] See, for example, this passage by Palomba (1995): "The decline in Italian fertility appears linked to (if not directly caused by) the high value attributed to motherhood. This does not mean an attitude of opposition to having children. On the contrary, the well-being of children and total devotion to their up bring are seen as the greatest values in life. The fall in the birth rate in Italy is taking place without any simultaneous increase in marriages without children or in cohabitation. The fall in the number of children per woman (..) may well be explained by the hypothesis of an 'improved quality' of children. Italian parents want to be better parents for a lower number of children".

[44] In order to facilitate the visual perception of the findings, the tables do no bear the numerical values of the parameters of the models, but only their algebraic sign and their statistical significance.

[45] Indeed, this research is obligatory as well as legitimate, since strictly cognitive mechanisms fail to provide satisfactory interpretations of social change.

[46] Riesman (1953) suggested the use of the metaphor of the gyroscope in order to indicate the deep layers of identity, which 'hold the course' in individual and collective behaviour.

[47] "The individual strives an action goal not only for the sake of its intrinsic value but also because its attainment implies positive information about the self in terms of self-ascribed competencies (..). Anticipatory self-reinforcement becomes a key mechanism for maintaining tolerance for the delay of gratifications regarding long-

term goals as well as persistence in the face of failures and obstacles (..). However the self-concept as a motivational resource also has its perils, when it comes to managing the implications of failures for negative perceptions of the self. Failure experiences produce negative information about the self and thus may threaten self-ascribed competencies, and even self-esteem in general. To prevent such damage to motivational and emotional resources the organism needs specifically adapted strategies to compensate for such negative implications of failure on the self" (Heckausen 1999).

[48] In other words: "cognitive dissonance is stipulated to arise when a person holds two or more 'cognitions' (including both beliefs and consciously held values as well as mental representations of the choices or behaviours of the subject; even emotions) that are inconsistent with one another. A and B are inconsistent when the presence of A induces an expectation of the absence of B. Dissonance involves physiological (painful) arousal. Dissonance reduction takes place by changing or blocking some of the dissonant cognitions, and sometimes by adding new ones. The process has to be thought of as unconscious" (Elster, 1999).

[49] "Such strategies typically are directed at the internal world rather than at the environment, and specifically target mental representations of expectations, goals and causal attributions (secondary control strategies)" (Heckausen, 1999).

[50] "Rules and normative systems cannot be violated, but their interpretation can be manipulated. Through the concept of manipulative interpretation, strategic action is reintroduced within the very fabric of normative action. The distinction between rules of varying manipulability is analogous to that between rules and values" (Rositi, 1986).

[51] This is not an isolated choice. In the past Lazarsfeld (1948), for example, argued that the putting off of the decision to vote (a change which apparently involves only the cognitive structure of decisions) is just one of the possible manifestations of a more general inability, among individuals exposed to contrasting pressures (often the effect of status inconsistency), to take any decision whatsoever (mechanism of logical-affective destructuring).

[52] From a formal point of view a contradiction is an instability located within a particular set of relations or references. From an affective point of view, it is discordance, i.e., the collision between two opposing dispositions. The paradox, on the other hand, is a contradiction, which derives from a correct deduction from coherent premises, and it is therefore more than a simple contradiction. It results from the meeting of two logical-affective systems of reference of the same order, each in itself valid, but irreconcilable with the other. Formally, unlike the simple contradiction, the paradox does not simply contain a discrepancy circumscribed within a dominant system of reference, but there are two ways of interpreting reality, which are equivalent but contradictory with respect to each other, and which enter into collision. Paradoxically opposed truths are - necessarily - true and false at the same time".

[53] "Evidently it is impossible to live at the same time in two different logical-affective systems of reference (since) each reference system is, at the same time, also

a system of values which motivates action. Opposite values are as barely compatible as opposite states of mind" (*ibid.*).

[54] "The psychological basis upon which the type of metropolitan individuality stands is the intensification of nervous life, which is the product of the rapid and uninterrupted succession of external and internal impressions (...) The metropolitan type creates an organ of defence against the uprooting with which he is threatened by his external environment: he reacts essentially with his intellect, as opposed to his whole set of feelings. Consequently, his reaction to phenomena is shifted onto that organ of the psyche which is least sensitive and furthest away from his deep layers of the personality (...) There is perhaps no psychic phenomenon which is so irreducibly reserved to the metropolis as that of being *blasé*, a character resulting from the rapid succession and dense concentration of nervous and contradictory stimuli, from which derives -or so we believe - the increase in metropolitan intellectualism" (Simmel, 1903).

[55] For how to read the table, see note 44.

[56] In the Friuli survey, the quality of primary relations in the early stages of life was explored by means of the following question: "Think back to your family life when you were a child (up to 7-8 years) and please indicate the extent to which the following statements describe your family experience: a) my parents gave me all the affection they could; b) I remember frequent disagreements between my parents; c) we often laughed and sang at home; d) my father was very much present in my life; e) every aspect of my childhood was marked by economic hardship". For each suggested scenario, the respondent had to declare the degree of correspondence (on a 5-point scale) with her own family experience. Items a) and d) were combined in an index of affective deprivation, items b) and c) in an index of lack of domestic harmony, while item 9 measured the living standards of the family of origin.

[57] The proper number of children falls drastically to about 1.5, from a regional average around 2.

[58] A child develops affective ties with its own figure of reference (generally the parent), which then always remain active, even between adult and adult; the experience in childhood of disturbed models of attachment can indeed either give rise in adult age to an excessive, anxious attachment, or to a total de-activation of the attachment. Bowlby's theory of disturbed imprinting belongs to larger family of theories of psycho-pathogenesis, which hinge on mechanisms of intergenerational transmission. These include the multigenerational theory of schizophrenia (Boszormenyi-Nagy and Spark, 1973), or the theory of parent-child fusion (Selvini Palazzoli, *et al.* 1975).

[59] Mechanisms in which the first stage (reaction to critical events or to unbearably contradictory situations) follows the logic of cognitive dissonance, while the second (extension of the reactive state to a destructured and non goal-oriented dispositional state) takes place in the phase of imprinting.

[60] "Dans le domaine éthique et moral, l'autonomie individuelle se traduit par la laicité, le refus de la morale institutionnelle et d'une éthique imposée, l'affirmation de la liberté de choix, le rejet du conformisme et une plus grande tolérance pour les modes de vie des autres" (Lesthaeghe and Moors, 1995).

REFERENCES

Adorno T.W. *et al.* (1973) *La personalità autoritaria*, Torino, Comunità.

Ainsworth, M.D.S. *et al.* (1978) *Patterns of Attachment*, Hillsdale, Erlbaum.

Allegra S. F., A. Rosina (2001) "Genitori e figli", *unpublished.*

Allport, G.W. (1935) "Attitudes", in C. Murchison (ed.) *A Handbook of Social Psychology*, Worcester, Clark University Press, 798-884.

Almond G., S. Verba (1963) *The civic culture. Political attitudes and democracy in five nations*, Princeton, Princeton University Press.

Altan C.T. (1986) *La nostra Italia*, Milano, Feltrinelli.

Anderson M. (1971) *Family Structure in Nineteenth Century Lancashire*, Cambridge, Cambridge University Press.

Anderson M. (1972) "Household structure and the industrial revolution; mid-nineteenth-century Preston in comparative perspective", in P. Laslett (1972 ed.), 215-235.

Anderson M. (1977) "The impact on the family relationships of the elderly of changes since Victorian times in governmental income-maintenance", in E. Shanas, M. Sussman (eds.) *Family, Bureaucracy and the Elderly*, Durham, Duke University Press.

Angeli A., A. Bellettini (1979) "Strutture familiari nella campagna bolognese a metà del'Ottocento", *Genus*, XXXV, 3-4, 155-172.

Arru A. (1990) "The distinguishing features of domestic service in Italy", *Journal of Family History*, 15, 4, 547-566.

Baali F., A. Wardi (1981) *Ibn Khaldun and Islamic Thought-Styles. A Social Perspective*, Boston, G.K. Hall & Co.

Bainton R.H. (1952) *The Reformation of the Sixteenth Century*, Boston, The Beacon Press.

Baizán P., A. Aassve, F.C. Billari (2001) *Cohabitation, marriage, first birth: The interrelationship of family formation events in Spain*, Max Plank Institute for Demographic Research.

Balbo L. (1976) *Stato di famiglia*, Milano, Etas.

Balbo L., P. May, G.A. Micheli (1990 eds.) *Vincoli e strategie nella vita quotidiana: una ricerca in Emilia-Romagna*, Milano, Franco Angeli.

Banfield E.C. (1958) *The moral basis of a backward society*, New York, The Free Press.

Baran P.A. (1957) *The Political Economy of Growth*. New York, Monthly Review Press.

Barbagli M. (1984) *Sotto lo stesso tetto: Mutamenti della famiglia in Italia dal XV al XX secolo*, Bologna, il Mulino.

Barbagli M. (1990) *Provando e riprovando*, Bologna, il Mulino.

Barbagli M. (1991) "Three Household Formation Systems in 18th and 19th Century Italy", in D.I. Kertzer, R.P. Saller (eds.) 279-301.
Barbagli M. (1997) "Family and kinship in Italy", in M. Gullestad, M. Segalen (eds.) 33-48.
Barbagli M., C. Saraceno (1997 eds.) *Lo stato delle famiglie in Italia*, Bologna, il Mulino.
Barbagli M., M. Castiglioni, G. Dalla Zuanna (2003), *Fare famiglia in Italia*, Bologna, il Mulino.
Becker G. (1981) *A treatise on the family*, Cambridge, Harvard University Press.
Bégeot F., J.A. Fernández-Cordón (1997) "La convergence démographique au-delà des différences nationales", in J. Commaille, F. de Singly (eds.) 33-60.
Bernardi F. (1999) *Donne fra famiglia e carriera*, Milano, Franco Angeli.
Bestard Camps J., J. Contreras Hernandez (1999) "Family, Kinship and Residence in Urban Catalonia: the Modernity of 'Pairalism'", in M. Gullestad, M. Segalen (eds.) 61-76.
Bettio F., P. Villa (1998) "A Mediterranean Perspective on the Break-Down of the Relationship Between Participation and Fertility", *Cambridge Journal of Economics*, 22, 2, 137-171.
Billari F.C., A. Rosina (2003) *Percorsi e tempi di transizione allo stato adulto e fecondità in ambito urbano*, in M. Breschi, M. Livi Bacci (eds.) 137-150.
Billari F.C., G.A. Micheli (1999) "Le scelte demografiche: la percezione dei costi e delle norme sociali", in L. Mauri, F. Billari (eds.) 163-190.
Billari F.C., G.A. Micheli (2001) "Social norms on agents' demographic events. Preliminaries for a (multi-) agent-based approach", in R. Conte, C. Dellarocas (eds.) *Social Order in Multiagent Systems*, Dordrecht, Kluwer, 201-221.
Billari F.C., H.P. Kohler, J.A. Ortega (2002) "The Emergence of Lowest-Low Fertility in Europe During the 1990s", *Population and Development Review*, 28, 4, 641-680.
Billari F.C., M. Castiglioni, F. Ongaro (1999b) *L'uscita dei giovani dalla famiglia di origine,* paper presented at the Conference "Le famiglie interrogano le politiche sociali", Session titled "Politiche amichevoli nei confronti delle nascite", Bologna, March 29-31.
Billari F.C., M. Castiglioni, T. Castro Martin, F. Michelin, F. Ongaro (2000) "Household and Union Formation in a Mediterranean Fashion: Italy and Spain", *FFS Flagship Conference* Brussels, 29-31 May.
Billari F.C., P. Manfredi, A. Valentini (1999a) *Becoming adult and its macro-demographic impact: Multistate stable population theory and an application to Italy*, paper prepared for session J of the European Population Conference, The Hague, 30 August - 3 September.
Bimbi F., S. La Mendola (1999) "Contratti di genere e modelli femminili e maschili della doppia presenza", in P. De Sandre *et al.* (eds.) 255-272.
Binswanger L. (1956) *Drei Formen Missgluckten Daseins*, Tubingen, Max Niemeyer Verlag.

Blumin S.M. (1977) "Rip Van Winkle's grandchildren: Family and household in the Hudson Valley, 1800-1860", in T.K. Hareven (ed.) 100-121.

Bongaarts J., G. Feeney (1998) "On the quantum and tempo of fertility", *Population and Development Review*, 24, 2, 271-291.

Boserup E. (1970) *Women's Role in Economic Development*, London, Allen & Unwin.

Boszormenyi-Nagy I., G.M. Spark (1973) *Invisible Loyalties. Reciprocity in Intergenerational Family Therapy*, New York, Harper and Row.

Bott E. (1957) *Family and Social Network. Roles, Norms and External Relationships in Ordinary Urban Families*, London, Tavistock.

Bott E. (1971) "Reconsiderations", in E. Bott (ed.) *Family and Social Network. Roles, Norms and External Relationships in Ordinary Urban Families* (2nd edition), London, Tavistock, 248-330.

Bourgeois-Pichat J. (1987) "La demografia e le altre scienze: interazione di metodo e di contenuto", in E. Sonnino *et al.* (eds.) *Demografia: scienza insegnamento professione*, Milano, Franco Angeli, 53-67.

Bowlby J. (1969) *Attachment and Loss*, vol. 1, New York, Basic Books.

Brandão M.S. (1994) *Terra, herença e família no noroeste de Portugal. O caso de Mosteiro no século XIX*, Porto, Edições Afrontamento.

Breschi M., M. Livi Bacci (2003 eds.) *La bassa fecondità italiana tra costrizioni economiche e cambio di valori*, Udine, Forum.

Brettell C.B. (1991) "Property, Kinship and Gender: a Mediterranean Perspective", in D.I. Kertzer, R.P. Saller (eds.) 381-395.

Brooke M.Z. (1998) *Le Play Engineer and Social Scientist*, New Brunswick, Transaction.

Brown G.W., T. Harris (1978) *Social Origins of Depression*, London, Tavistock.

Burch T., B. Matthews (1987) "Household formation in developed societies", *Population and Development Review*, 13, 3, 495-511.

Burguière A. (1986) "Une géographie des formes familiales", in A. Burguière *et al.* (eds.) vol. 2: *Le choc des modernités*, 25-92.

Burguière A. (1997) "Les fondements historiques des structures familiales", in J. Commaille, F. Singly (eds.) 137-155.

Burguière A., C. Klapisch-Zuber, M. Segalen, F. Zonabend (1986 eds.) *Histoire de la famille*, Paris, Armand Colin.

Buzzi A. (1997) "Rischio, reversibilità, sfiducia negli altri, disagio", in C. Buzzi, A. Cavalli, A. de Lillo (eds.) 87-102.

Buzzi C., A. Cavalli, A. de Lillo (1997 eds.) *Giovani verso il Duemila*, Bologna, il Mulino.

Caldwell J.C. (1982) *The Theory of Fertility Decline*, New York, Academic Press.

Calot G. (1984) "Une notion intéressante: l'effectif moyen des générations soumises au risque: présentation méthodologique", *Population*, 39, 6, 949-976.

Cantisani G., G. Dalla Zuanna (1999) "The new fertility transition in Europe. Have the gaps been bridged?", *Proceedings of the Third African Population Conference*, Durban, South Africa. December. UAPS-IUSSP, vol. 3, 503-520.

Carroll E.V. (1988) "Share and share alike: Inheritance patterns in two Illinois farm communities", *Journal of Family History*, 13, 2, 219-237.
Casacchia O., G. Dalla Zuanna (1999) "Trasmissione intergenerazionale di fecondità e interrelazioni con la mobilità sociale", in P. De Sandre *et al.* (eds.) 555-567.
Castiglioni M., G. Dalla Zuanna (1994) "Innovation and tradition: reproductive and marital behaviour in Italy in the 1970s and 1980s", *European Journal of Population*, 10, 107-141.
Castiglioni M., G. Dalla Zuanna (1995) "Some indications on reproductive behaviour of never married women in Italy", *Contribution of Italian scholars to the European Population Conference*, 5-21, Rome, IRP-CNR.
Cesareo V., R. Cipriani, F. Garelli, C. Lanzetti, G. Rovati (1995) *La religiosità in Italia*, Milano, Mondadori.
Chacón Jiménez F. (1983) "Introducción a la historia de la familia española: el ejemplo de Murcia and Orihuela (siglos XVII-XIX)", *Cuadernos de Historia*, 10, 235-266.
Chudacoff H.P., T.K. Hareven (1979) "From the empty nest to family dissolution: Life course transitions into old age", *Journal of Family History*, 4, 1, 69-83.
Cigno A., F.C. Rosati (1992) "The effect of financial markets and social security on saving and fertility behaviour in Italy", *Journal of Population Economics*, 5, 3, 319-341.
Ciompi L. (1982) *Affektlogik. Uber die Struktur der Psyche und ihre Entwicklung*, Stuttgart, Ernst Klett.
Clerici R. (1999) "Pratica religiosa e scelte riproduttive", in P. De Sandre *et al.* (eds.) 295-310.
Coale A.J., S.C. Watkins (1986 eds.). *The Decline of Fertility in Europe*, Princeton, Princeton University Press.
Commaille J., F. Singly (1997 eds.) *La question familiale en Europe*, Paris, L'Harmattan.
Crosbie P.V. (1984) "Rationality and Models of Reproductive Decision-Making", in W. Penn Handwerker (ed.) *Culture and Reproduction* Boulder, Westview Press, 30-58.
Cuvillier J.P. (1986) "l'«Urfamilie» germanique: peuple, clan, maison", in A. Burguière *et al.* (eds.) *Histoire de la Famille*, vol. 1: *Mondes lointains, mondes anciens*, Paris, Armand Colin, 293-331,
Da Molin G. (1990a) *La famiglia nel passato. Strutture familiari nel Regno di Napili in età moderna*, Bari, Cacucci Editore.
Da Molin G. (1990b) "Family forms and domestic service in southern Italy from the seventeenth to the nineteenth centuries", *Journal of Family History*, 15, 4, 503-527.
Dalla Zuanna G. (1995) "Meglio soli. Famiglia e natalità in Italia", *il Mulino*, XLIV, 357, 111-118.
Dalla Zuanna G. M. Atoh, M. Castiglioni, K. Koshima (1998) "Late marriage among young people: the case of Italy and Japan", *Genus*, LIV, 3-4, 187-232.

Dalla Zuanna G., A. Righi (1999) *Nascere nelle cento Italie*, Roma, Istat Collana Argomenti, 18.
Dalla Zuanna G., C. Crisafulli (2001) *Come interpretare il rialzo di fecondità in Italia negli ultimi anni del XX secolo?*, Workshop *Venus*, Firenze, 8-9 November 2001.
Davitz J.R. (1969) *The language of emotions*, New York, Academic Press.
de Jong-Gierveld J., H. van Solinge (1995) *Ageing and Its Consequences for the Socio-Medical System*, Strasbourg, Council of Europe, Directorate of Social and Economic Affairs (*Population Studies* no. 29).
De Masi D. (1976 ed.) *Le basi morali di una società arretrata*, Bologna, il Mulino.
De Rose A., F. Racioppi (2001) *Describing and explaining differences in fertility among the European countries: a multilevel approach*, European Population Conference, Session A.2, , The Hague, 30 August - 3 September.
De Rosnay J. (1975) *Il macroscopio*, Bari, Dedalo.
De Sandre I., G. Dalla Zuanna (1999) "Correnti socioculturali e comportamento riproduttivo", in P. De Sandre *et al.* (eds.) 273-294.
De Sandre P. (2000) "Patterns of fertility in Italy and factors of its decline", *Genus*, LVI, 1-2, 19-54.
De Sandre P., A. Pinnelli, A. Santini (1999 eds.) *Nuzialità e fecondità in trasformazione: percorsi e fattori del cambiamento*, Bologna, il Mulino.
De Sandre P., F. Ongaro, R. Rettaroli, S. Salvini (1997) *Matrimonio e figli: tra rinvio e rinuncia*, Bologna, il Mulino.
De Santis G. (1997) *Economia e demografia*, Bologna, il Mulino.
De Santis G., A. Righi (1997) "Il costo dei figli per le famiglie italiane", *Polis*, XI, 1, 29-50.
Delgado Pérez M., M. Livi Bacci (1992) "Fertility in Italy and Spain: The lowest in the world", *Family Planning Perspective*, 24, 162-167.
Delille G. (1985) *Famille et propiété dans le Royaume de Naples (XV-XIX siècles)*, Rome, Edition de l'Ecole des Hautes Etudes en Sciences Sociales, trad. it. (1988) *Famiglia e proprietà nel regno di Napoli*, Torino, Einaudi.
Di Giulio P., R. Lesthaeghe, H. Moors, A. Pinnelli (2000) "Fertility, tempo and quantum: an empirical test of major theories with data from four FFS countries", in A. Caselli, J. Vallin, G. Wunsch (eds.) *Démographie: analyse et synthèse, Actes du Séminaire de Roma – Napoli – Pozzuoli, vol. 2*, 31-48, Materiali di studi e ricerche (numero speciale), Dipartimento di Scienze Demografiche, Roma.
Ditch J., H. Barnes, J. Bradshaw (1996), *A synthesis of national family policies 1995*, European Observatory on National Family Policies, European Commission Social Policy Research Unit, York.
Doveri A. (1993) "«Padre che ha figliuoli grandi fuor li mandi». Una prima valutazione sulla diffusione e sul ruolo dei «garzoni» nelle campagne pisane dei secoli XVII e XVIII", in Società Italiana di Demografia Storica (ed.) *La popolazione delle campagne italiane in età moderna*, Bologna, CLUEB, 426-229.

Drake M. (1969) *Population and Society in Norway*, Cambridge, Cambridge University Press.
Dubert G.I. (1992) *Historia de la familia en Galicia durante la época moderna, 1550-1830 (Estructura, modelos hereditarios y conflictividad)*, A Coruña, Ediciós do Castro.
Durkheim E. (1895) *Les règles de la méthode sociologique*, Paris, Alcan.
Easterlin R. (1976) "The conflict between aspirations and resources", *Population and Development Review*, 2, 3, 417-426.
Ehmer J. (1998) "House and the stem family in Austria", paper presented for session C18 *The stem-family in Eurasian perspective, as domestic unit of production and reproduction, facing economic and social change* of the *Twelfth International Economic History Congress*, Seville.
Eibl Eibesfeld I. (1983) *Etologia della Guerra*, Torino, Boringhieri.
Ekart-Jaffé O. (1994) "Chiffrer une évolution du cout de l'enfant?", *Population*, 6, 1389-1418.
Elder G.H., A. Caspi (1988), *Studying Lives in a Changing Society*, Henry Murray Lecture Series, Michigan State University.
Elster J. (1999) *Alchemies of the Mind. Rationality and the Emotions*, Cambridge, Cambridge University Press.
Esping-Andersen G. (1990) *The three worlds of welfare capitalism*. Princeton, Princeton University Press.
Esping-Andersen G. (1999) *Social foundations of post-industrial economy*. Oxford, Oxford University Press.
Etzioni A. (1988) "Normative-affective factors: toward a new decision-making model", *Journal of Economic-Psychology*, 9, 125-150.
European Observatory on National Family Policies (1998) *A Synthesis of National Family Policies in 1996*, European Commission.
Fawcett J.T. (1970) *Psychology and Population. Behavioral Research Issues in Fertility and Family Planning*, New York, The Population Council.
Federkeil G. (1997) "The Federal Republic of Germany: Polarisation of Family Structure", in F.X. Kaufmann *et al.* (eds.) *Family Life and Family Policies in Europe*, Oxford, Clarendon Press, 77-112.
Fernández Cordón J.A. (1997) "Youth residential independence and autonomy. A comparative study", *Journal of Family Issues*, 6, 568-575.
Fernández Cordón J.A., C. Tobío-Soler. In press. "Las familias monoparentales en España", *Revista Española de Investigación Sociológica (REIS)*.
Festinger L. (1957) *A Theory of Cognitive Dissonance*, Stanford, Stanford University Press.
Fortis M. (1998) *Il made in Italy*, Bologna, il Mulino.
Fraboni R., A. Rosina, C. Freguja, L.L. Sabbadini (2001) *Transition to Fatherhood in Italy*, paper presented at General Conference IUSSP, Salvador de Bahia, Brazil, Session 22, 18-24 August.
Frejka T., G. Calot (2001) "Cohort reproductive patterns in low-fertility countries", *Population and Development Review*, 27, 1, 103-132.

Gabrieli F. (1930) "Il concetto della 'Asabiyyah' nel pensiero storico di Ibn Haldun", *Atti Reale Accademia delle Scienze di Torino*, 65, 473-512.

Garelli F. (1984) *La generazione della vita quotidiana*, Bologna, il Mulino.

Garrido Arce E. (1992) "La imposible igualdad. Familia y estrategias hereditarias en la Huerta de Valencia a mediados del siglo XVIII", *Boletín de la Asociación de Demografía Histórica*, 10, 3, 83-104.

Gaunt D. (1983) "The property and kin relationships of retired farmers in northern and central Europe" in R. Wall et al. (eds.) 249-279.

Gauthier A.H. (2002) Family policies in industrialized countries: Is there convergence?, *Population*, 57, 3, 447-474.

Giddens A. (1992) *The Transformation of Intimacy*, Cambridge, Polity Press.

Gierveld J., T. van Tilburg, L. Lecchini (1995) *Socioeconomic Resources, Living Arrangements and the Social Network as Determinants of Well-Being among Older Adults*, Paper presented at the EAPS Population Conference, Milano, Italy.

Ginsborg P. (1989) *Storia d'Italia dal dopoguerra a oggi*, Torino, Einaudi, (English translation: *A History of Contemporary Italy: Society and Politics: 1943-1980*, Penguin Books).

Ginsborg P. (1994) "Familismo", in P. Ginsborg (ed.) *Stato dell'Italia*, Milano, il Saggiatore, 78-82.

Ginsborg P. (1998) *L'Italia del tempo presente*, Einaudi, Torino. (English translation: *A History of Contemporary Italy, Society and Politics 1980-2001*, Penguin Books).

Glasco L.A. (1977) "The life cycles and household structure of American ethnic groups: Irish, Germans and native-born whites in Buffalo, New York, 1855", in T.K. Hareven (ed.) 122-143.

Goldstein J.R., W. Lutz, M.R. Testa (2003) *The Emergence of Sub-Replacement Family Size Ideals in Europe*, Paper presented at the Annual Meeting of the Population Association of America, Minneapolis.

Golini A. (1994 ed.), *Tendenze demografiche e politiche per la popolazione*, Bologna, il Mulino.

Golini A. (1997) *Levels and trends of fertility in Italy. Are they desirable and/or sustainable?*, Expert Group Meeting on "Below-replacement fertility", organised by the Population Division of the Department of Economic and Social Affairs, United Nations, New York, 4-6 November, preliminarily version.

Golini A. (1998) "How Long Can Fertility Be? An Empirical Exploration", *Population and Development Review*, 24, 1, 59-73.

Goody J. (1983) *The Development of the Family and Marriage in Europe*, Cambridge, Cambridge University Press (trad. it. *Famiglia e matrimonio in Europa*, Roma-Bari, Laterza).

Goody J. (1996a) "Comparing family systems in Europe and Asia: Are there different sets of rules?" *Population and Development Review*, 22, 1, 1-20.

Goody J. (1996b) *The East in the West*, Cambridge, Cambridge University Press.

Granovetter M. (1973) "The Strength of Weak Ties", *American Journal of Sociology*, 78, 1360-1380.
Gribaudi G. (1994) "Il paradigma del familismo amorale", in P. Macry, A. Massafra (eds.) *Fra storia e storiografia*, Bologna, il Mulino.
Gubert R. (2000 ed.) *La via italiana alla postmodernità. Verso una nuova architettura dei valori*, Milano, Franco Angeli.
Guicciardini F. (1970) *Opere* (Vol. I), Torino, Utet.
Guichard P. (1977) *Structures sociales 'orientales' et 'occidentales' dan l'Espagne musulmane,* Paris, Mouton.
Gullestad G. (1997) "From 'Being of Use' to 'Finding Oneself', Dilemmas of Value Transmission Between the Generations in Norway", in M. Gullestad, M. Segalen (eds.) 202-218.
Gullestad M., M. Segalen (1997 eds.) *Family and Kinship in Europe*, London, Pinter.
Hajnal J. (1965) "European marriage patterns in perspective", in D.V. Glass, D.E.C. Eversley (eds.) *Population in History'*, Londres, Edward Arnold, 101-146.
Hajnal J. (1982) "Two kinds of preindustrial household formation system", *Population and Development Review*, 8, 3, 449-494.
Hantrais L., M.T. Letablier (1996) *Familles, travail et politiques familiales en Europe*. Cahiers du Centre d'Études de l'Emploi, Paris, Press Universitaires de France.
Hareven T.K. (1977 ed.) *Family and Kin in Urban Communities, 1700-1930*, New York-London, New Viewpoints.
Harris C.C. (1969) *The Family*, London, Allen & Unwin.
Heckausen J. (1999) *Developmental Regulation in Adulthood. Age-normative and sociostructural constraints as adaptive challenges*, Cambridge, Cambridge University Press.
Herlihy D., C. Klapisch-Zuber (1978) *Les Toscans et leurs familles: Une étude du catasto florentin de 1427*, Paris, Fondation Nationale des Sciences Politiques.
Hill D., M. Hill (1976) "Older children and splitting off", in G. Duncan G., J. Morgan (eds.) *Five thousand American families – Patterns of economic progress*, Vol. IV, 117-154, Institute for Social Research, Ann Arbor.
Hionidou V. (1995) "Nuptiality patterns and household structure on the Greek island of Mykonos, 1849-1959", *Journal of Family History*, 20, 1, 67-102.
Hoem J.M. (1986) "The impact of education on modern family union initiation", *European Journal of Population*, 2, 113-133.
Holdsworth C. (1998) "Leaving Home in Spain: a Regional Analysis", *International Journal of Population Geography*, 4, 341-360.
Hollinger F., M. Haller (1990) "Kinship and Social Networks in Modern Societies: a Cross-Cultural Comparison Among Seven Nations", *European Sociological Review*, 6, 2, 103-124.
Hufton O.H. (1974) *The Poor of Eighteenth-Century France 1750-1789*, Oxford, Clarendon Press.

Ibn Khaldoun (1965) *Les textes sociologiques et économiques de la Mouqaddima*, Paris, Marcel Rivière.

Iceland Statististical Bureau (1975) H.O. Hansen, *Population Census 1729 in Three Counties* (Statistics of Iceland II, 59), Reykjavik.

Inglehart R. (1990) *Culture shift in advanced industrial society*, Princeton, Princeton University Press.

Inglehart R. (1997) *Modernisation and postmodernisation*, Princeton, Princeton University Press.

IreR (1991) *Social Survey in Lombardia. Contributi alla formazione della domanda sociale*, Milano, Franco Angeli.

Irp - National Institute for Population Research (1999) *Giovani che non lasciano il nido*, IRP working paper, 1/99.

Isen A.M. (1984) "Towards understanding the role of affect in cognition", in R.S. Wyer, T.K. Srull (eds.) *Handbook of social cognition*. Hillsdale, Erlbaum.

Istat (1998) *La fecondità regionale nel 1995*, Roma, Collana Informazioni, 97.

Istituto Nazionale di Statistica (1997) *La fecondità nelle regioni italiane. Analisi per coorti. Anni 1952-1993*, Roma, Istat Informazioni, n. 35.

Jani-Le Bris H. (1993) *Cuidado familiar de las personas de edad avanzada en la Comunidad Europea*, Luxembourg, Oficina de Publicaciones Oficiales de las Comunidades Europeas (Fundación europea para la mejora de las condiciones de vida y de trabajo).

Johansen H.C. (1975) *Befolkningsudvikling og Familiestruktur i det 18 Arhundrede*, Odense.

Jowell R., S. Withers Poon, L. Brook (1989 eds.) *British social attitudes*, London.

Kertzer D.I. (1989) "The joint family household revisited: Demographic constraints and household complexity in the European past", *Journal of Family History*, 14, 1, 1-15.

Kertzer D.I. (1995) "Toward a historical demography of aging", in D. Kertzer, P. Laslett (eds.) *Aging in the Past. Demography, Society, and Old Age*, Berkeley, University of California Press, 363-383.

Kertzer D.I., P. Laslett (1995 eds.) *Aging in the Past. Demography, Society and Old Age*, Berkeley, University of California Press.

Kertzer D.I., R.P. Saller (1991 eds.) *The Family in Italy from Antiquity to the Present*, New Haven, Yale University Press.

Khoo S.E., P. McDonald, D. Giorgas, B. Birrel (2002), *Second Generation Australians*, Report for the Department of Immigration and Multicultural and Indigenous Affairs, Dimia, Canberra, Australia.

Kiernan K. (2002) "The state of European unions: an analysis of partnership formation and dissolution", in M. Macura, G. Beets (eds.) 57-76

Klapisch C. (1972) "Household and family in Tuscany in 1427", in P. Laslett (ed.) 267-282.

Klapisch C., M. Demonet (1972) "'A uno vino e uno pane.' La famille rurale toscane au début du Xve siècle." *Annales, E.S.C.*, 27, 4-5, 873-901.

Kohler H.P. (2001) *Fertility and Social Interactions: An Economic Perspective*, Oxford, Oxford University Press.
Kohler H.P., A. Skyttle, K. Christensen (2001) *The Age at First Birth and Completed Fertility Reconsidered: Findings from a Sample of Identical Twins*, Max Plank Institute for Demographic Research, Rostock, WP 2001-006.
Krishnan V. (1990). "A causal approach to the study of fertility and familism", *Social Biology*, 37, 1-2, 59-68.
Kuijsten A.C. (1996) "Changing family patterns in Europe: A case of divergence?" *European Journal of Population*, 12, 115-143.
Kussmaul A. (1981) *Servants in Husbandry in Early Modern England*, Cambridge, Cambridge University Press.
Lanza García R. (1991) *La población y el crecimiento económico de Cantabria en el Antiguo Régimen*, Madrid, Universidad Autónoma de Madrid - Universidad de Cantabria.
Laslett P. (1972 ed.) *Household and Family in Past Time*, Cambridge, Cambridge University Press.
Laslett P. (1977a) *Family Life and Illicit Love in Earlier Generations*, Cambridge, Cambridge University Press.
Laslett P. (1977b) "Characteristics of the western family considered over time", *Journal of Family History*, 2, 2, 89-115.
Laslett P. (1983) "Family and Households as Work Group and Kin Group: Areas of Traditional Europe Compared", in R. Wall *et al.* (eds.) 513-587.
Laslett P. (1984) "The significance of the past in the study of ageing", *Ageing and Society*, 4, 4, 379-389.
Laslett P. (1988) "Family, kinship and collectivity as systems of support in pre-industrial Europe: a consideration of the 'nuclear hardship' hypothesis", *Continuity and Change*, 3, 2, 153-176
Laslett P. (1989) *A Fresh Map of Life. The Emergence of the Third age*, London, Weidenfeld and Nicolson.
Lazarsfeld P.K. (1948) *The people's choice* (2nd edition), New York, Columbia University Press.
Le Bras H. (1999) "Fertility: the Condition of Self-Perpetuation. Differing Trends in Europe", in M. Gullestad, M. Segalen (eds.) 14-32.
Le Play F. (1855) *Les ouvriers européens* (2nd edition 1877-79), Tours, Alfred Mame et fils.
Le Play F. (1871) *L'organisation de la famille selon le vrai modèle signalé par l'histoire de toutes les races et de tous les temps* (3rd edition 1884), Tours , Alfred Mame et fils.
Lebrun F. (1986) "Le contrôle de la famille par les églises et par les états", in A. Burguière *et al.*, vol. 2, *Le choc des modernités*, 94-110.
Leibenstein H. (1981) "Economic decision, theory and human behaviour: a speculative essay", *Population and Development Theory*, 7, 3, 381-400.
Lesthaeghe R. (1991) *The Second Demographic Transition in Western Countries: an Interpretation*, IPD–WP, 2, Brussel, Vrije Universiteit Brussel.

Lesthaeghe R. (1995) "The second demographic transition: an interpretation", in K. O. Mason, A. M. Jensen (eds.) 17-62.

Lesthaeghe R. (1998) "On theory development and applications to the study of family formation", *Population and Development Review*, 24, 1, 1-14.

Lesthaeghe R. (1999) "Is low fertility a temporary phenomenon in the EU?" *Population and Development Review*, 25, 1.

Lesthaeghe R., D. Meekers (1986) "Value changes and the dimensions of familism in the European Community", *European Journal of Population*, 2, 3-4, 225-68.

Lesthaeghe R., G. Moors (1994) "Expliquer la diversité des formes familiales et domestiques: théories économiques et dimensions culturelles", *Population*, 49, 6, 1503-1526.

Lesthaeghe R., G. Moors (1995) "Is there a New Conservatism That Will Bring Back the Old Family? Ideational Trends and Stages of Family Formation in Germany, France, Belgium and the Netherlands 1981-1990", *Proceedings of EAPS Population Conference*, Milano, vol. I, 225-265.

Lesthaeghe R., P. Willems (1999) "Is Low Fertility a Temporary Phenomenon in the European Union?", *Population and Development Review*, 25, 2, 211-228.

Liefbroer A. (1998) *Life-planning during young adulthood: do intentions about the timing of family-life events come true?*, paper presented at the European Population Conference, The Hague, the Netherlands, 30 August - 3 September

Lilienfeld A.M., D.E. Lilienfeld (1980) *Foundations of Epidemiology*, New York, Oxford University.

Linton R. (1936) *The Study of Man*, New York, Appleton Century Crofts.

Litwak E., I. Szelenyi (1969) "Primary Group Structures and their Functions: Kin, Neighbours and Friends", *American Sociological Review*, 34, 465-481.

Livi Bacci M. (1977) *A History of Italian Fertility during the Last Two Centuries*, Princeton, Princeton University Press.

Livi Bacci M. (1997) "Abbondanza e scarsità. Le popolazioni d'Italia e d'Europa al passaggio del millennio", *il Mulino*, XLVI, 374, 993-1009.

Livi Bacci M., S. Salvini (2000) "Trop de famille et trop peu d'enfants: la fécondité en Italie depuis 1960", *Les Cahiers québécois de démographie*, 29, 2.

MacFarlane A. (1978) *The Origins of English Individualism*, Oxford, Basil Blackwell.

Macura M., G. Beets (2002 eds.) *Dynamics of fertility and partnership in Europe. Insights and lessons from comparative research. Volume I.* United Nations.

Macura M., G. Beets, M. Burkimsher (2002) "Fertility and partnership: why the FFS and what did we learn from it?", in M. Macura, G. Beets (eds.).

Maslow A. (1954) *Motivation and personality*, New York, Harper and Row.

Mason K.O., A.M. Jensen (1995 eds.) *Gender and Family Change in Industrialized Societies*, Oxford, Clarendon Press.

Mauri L., F.C. Billari (1999 eds.) *Generazioni di donne a confronto*, Milano, Franco Angeli.

Mauri L., G.A. Micheli (1992 eds.) *Vita di famiglia. Social Survey in Veneto*, Milano, Franco Angeli.

McDonald J., A. Rosina (2001) "Mixture Modelling of Recurrent Event Times with Long-Term Survivors: Analysis of Hutterite Birth Intervals", *Statistical Methods & Applications*, 10.

McIntosh M. (1984) "Servants and the household unit in an Elizabethan English community", *Journal of Family History*, 9, 1, 3-23.

Mendes J.M.A. (1995) *Trás-os-Montes nos fins do século XVIII segundo um manuscrito de 1796*. Lisbon, Fundação Calouste Gulbenkian/Junta Nacional de Investigação Científica e Tecnológica (2nd edition), Quadro XIII, 566 - A desdobrável.

Merton R.K. (1957) *Social Theory and Social Structure*, Glencoe, The Free Press.

Micheli G.A. (1991) *Generazioni. Il comportamento procreativo nell'ottica di una demografia comprendente*, Milano, Franco Angeli.

Micheli G.A. (1995 ed.) *La società del figlio assente. Voci a confronto sulla seconda transizione demografica*, Milano, Franco Angeli.

Micheli G.A. (1996) "New Patterns of Family Formation in Italy. Which Tools for Which Interpretations", *Genus*, 52, 1-2, 15-52.

Micheli G.A. (1999a) *Effetto generazione. Cinquant'anni di trasformazioni demografiche in Italia dal dopoguerra a oggi*, Roma, Carocci.

Micheli G.A. (1999b) "Come e quando si fanno (e si disfanno) i patti sociali e generazionali", in L. Mauri, F. Billari (eds.) 191-214.

Micheli G.A., F.C. Billari (1998) "Le forme-famiglia", in P. Natale (ed.) *L'Italia al macroscopio*, Milano, Feltrinelli, 134-157.

Mikelarena Peña F. (1995) *Demografía y familia en la Navarra tradicional*, Pamplona, Institución Príncipe de Viana.

Mitterauer M., R. Sieder (1977) *The European Family. Patriarchy to Partnership from the Middle Ages to the Present*, Chicago, University of Chicago Press.

Modell J., T.K. Hareven (1977) "Urbanization and the malleable household, An examination of boarding and lodging in American families", in T.K. Hareven (ed.) 164-186.

Moors H. (1990) *Attitudes toward demographic trends and population policy acceptance: Italy and the Netherlands in a comparative perspective*, IRP - Italian Institute for Population Research, Working Paper, 2, Roma.

Mutti A. (1994) "I sentieri dello sviluppo", *Rassegna Italiana di Sociologia*, 35, 1, 109-119.

Nazareth J.M., F. Sousa (1981) "Salvaterra de Magos nos finais do século XVIII: aspectos sociodemográficos", *Análise Social*, XVII, 66, 315-412.

Nazareth J.M., F. Sousa (1983) "A demografia portuguesa em finais do Antigo Regime – aspectos sociodemográficos de Coruche", *Cadernos da Revista de História Económica e Social*, 4, (Lisboa, Livraria Sá da Costa Editora), 3-121.

Nowlis V, H. Nowlis (1956) "The description and analysis of mood" *Annals of the New York Academy of Psychology*, 65, 345-355.

Núñez C.E. (1997) "La educación como fuente del crecimiento", *Papeles de Economía*, 73, 213-242.

Oakley A. (1984) *The Captured Womb*, Oxford, Basil Blackwell.

Osswald H. (1990) "Dowry, norms and household formation. A case study from North Portugal", *Journal of Family History*, 15, 2, 201-224.
Palomba R. (1995) "Italy, the Invisible Change", in R. Palomba, H. Moors (eds.) *Population, Family and Welfare. A Comparative Survey of European Attitudes*, Oxford, Clarendon Press, 158-176.
Palomba R., L.L. Sabbadini (1993) "Female Life Strategies: the Way of Compromise", *Proceedings of the XXIII IUSSP General Conference, Montréal*, vol. 2, Liège, IUSSP, 219-231.
Parsons T., E.A. Shils (1951) *Toward a General Theory of Action. Theoretical Foundations for the Social Sciences*. Harvard, Harvard University Press.
Pérez García J.M. (1988) "La familia campesina en la Huerta de Valencia durante el siglo XVIII", *Boletín de la Asociación de Demografía Histórica*, 6, 2, 5-28.
Pérez-Fuentes H.P. (1993) *Vivir y morir en las minas. Estrategias familiares y relaciones de género en la primera industrialización vizcaína, 1877-1913*, Bilbao, Universidad del País Vasco.
Pérez-Fuentes H.P., A. Pareja Alonso (1997) "Ageing alone or in a family: the case of Bilbao, 1825-1935", *Continuity and Change*, 12, 1, 79-102.
Perrin N., M. Poulain (2002) *Les Italiens en Belgique*. Sybidi Papers, 28, Louvain-la-Neuve.
Piazza M. (1994) "Introduzione", in S. Biadene *et al.*, *Modelli territoriali e differenze di genere*, Milano, Feltrinelli.
Pinnelli A. (1995) "Woman's Condition, Low Fertility, and Emerging Union Patterns in Europe", in K.O. Mason, A.M. Jensen (eds).
Pinnelli A., A. De Rose, P. Di Giulio, A. Rosina (2002) "Interrelationships between partnership and fertility Behaviour", in M. Macura, G. Beets (eds.).
Piore M.J., C.F. Sabel (1984) *The Second Industrial Divide: Possibilities for Prosperity*, New York, Basic Books.
Pisati M. (2000) "La frequenza alla messa festiva", *Polis*, XIV, 1.
Polanyi K. (1944) *The Great Transformation. The Political and Economic Origins of Our Time*, New York, Rinehart.
Pooley C.G., J. Turnbull (1997) "Leaving home: the experience of migration from the parental household in Britain since c. 1770", *Journal of Family History*, 22, 4, 390-424.
Putnam R. (1993) *Making Democracy Works*, Princeton, Princeton University Press.
Reher D.S. (1988a) *Familia, población y sociedad en la provincia de Cuenca, 1700-1970*, Madrid, Siglo XXI-Centro de Investigaciones Sociológicas.
Reher D.S. (1988b) "Household and family on the Castilian meseta: The province of Cuenca from 1750 to 1970", *Journal of Family History*, 13, 1, 59-74.
Reher D.S. (1997) *Perspectives on the Family in Spain, Past and Present*, Oxford, Oxford University Press.
Reher D.S. (1998) "Mercado de trabajo y empleo en España durante el siglo actual: perspectivas nuevas sobre un tema tradicional", paper given at the Seminar on *Raíces históricas de los problemas de la Economía actual* organized by the *Real Academia de Ciencias Morales y Políticas*, Madrid.

Reher D.S., M.N. Pombo, B. Nogueras (1993) *España a la luz del Censo de 1887*, Madrid, Instituto Nacional de Estadística.
Reher D.S.. (1990) *Town and Country in Preindustrial Spain. Cuenca, 1550-1870*, Cambridge, Cambridge University Press, 1990.
Riesman D. (1953) *The Lonely Crowd. A Study of the Changing American Character*, New York, Doubleday.
Rokeach M. (1973) *The Nature of Human Values*, New York, The Free Press.
Rosina A. (1999) " Analisi delle storie riproduttive con un modello di mistura multiepisodio", in P. De Sandre *et al.* (eds.) 569-588.
Rosina A. (2001) *Quest'unione libera non s'ha da fare*, Workshop *Venus*, Firenze, 8-9 November 2001.
Rosina A., F.C. Billari (2003) "Flessibilità all'entrata in unione: i precursori del cambiamento", in M. Breschi, M. Livi Bacci (eds.) 127-136.
Rosina A., R. Fraboni, L.L. Sabbadini (2003) "Diventare uomini e donne in Italia", in A. Pinnelli *et al.* (eds.) *Genere e demografia in Europa*, Bologna, il Mulino.
Rositi F. (1986) "Tipi e dimensioni dei sistemi normativi", *Rassegna Italiana di Sociologia*, 27, 3, 347-367.
Roussel L. (1992) "La famille en Europe occidentale: Divergences et convergences", *Population*, 47, 1, 133-152.
Rowland R. (1988) "Sistemas matrimoniales en la Península Ibérica (siglos XVI-XIX). Una perspectiva regional", in V. Pérez Moreda, D.S. Reher (eds.) *Demografía Histórica en España*, Madrid, Ediciones El Arquero, 72-137.
Russell J.A. (1980) "A circumflex model of affect", *Journal of Personality and Social Psychology*, 39, 1161-1178.
Sahlins M. (1972) *Stone Age Economics*, Chicago, Aldine.
Santini A. (1995) *Continuità e discontinuità nel comportamento riproduttivo delle donne italiane nel dopoguerra: tendenze generali della fecondità delle coorti nelle ripartizioni tra il 1952 e il 1991*, WP, Dipartimento Statistico, Università degli Studi di Firenze.
Saraceno C. (1994) "The ambivalent familism of Italian welfare state", *Social Politics*, 1, 60-82.
Saraceno C. (1998) *Mutamenti della famiglia e politiche sociali in Italia*, Bologna, il Mulino.
Schellekens J. (1991) "Determinants of marriage patterns among agricultural laborers in two eighteenth century Dutch villages", *Journal of Family History*, 16, 2, 139-155.
Schmidtbauer P. (1983) "The changing household: Austrian household structure from the seventeenth to the early twentieth century", in R. Wall *et al.* (eds.) 347-378.
Schmitt C. (1963) *Theorie des Partisanen*, Berlin, Duncker und Humblot.
Schneider J., P. Schneider (1976) *Culture and Political Economy in Western Sicily*. New York, Academic Press.

Schoenmaeckers A.C., E. Lodewijckx (1999) "Demographic Behaviour in Europe: Some Results from FFS Country Reports and Suggestions for Further Research", *European Journal of Population*, 15, 207-240.

Schultheis F. (1999) "The Missing Link: Family Memory and Identity in Germany", in M. Gullestad, M. Segalen (eds.) 49-60.

Schumpeter J.A. (1943) *Capitalism, Socialism and Democracy*, London, Unwin.

Scott A.S.V. (1998) *Famílias, formas de união e reprodução social nor noroeste português*, unpublished doctoral dissertation, European University Institute, Florence.

Selvini Palazzoli M. *et al.* (1975) *Paradosso e controparadosso*, Milano, Feltrinelli.

Simmel G. (1903) "Die Grosstaedte und das Geistesleben", *Jahrbuch der Gehe-Stiftung*, IX.

Simmel G. (1907) *Philosophie des Gelde* (2nd edition), Leipzig-Berlin, Duncker und Humblot.

Smith J. (1984) "Widowhood and ageing in traditional English society", *Ageing and Society*, 4, 4, 429-449.

Smith R.M. (1981) "The people of Tuscany and their families in the fifteenth century: Medieval or Mediterranean", *Journal of Family History*, 6, 1, 107-128.

Smith R.M. (1984) "The structural dependence of the elderly as a recent development: Some skeptical historical thoughts", *Ageing and Society*, 4, 4, 409-428.

Strather M. (1992) *After Nature. English Kinship in the Late Twentieth Century*, Cambridge, Cambridge University Press.

Sumner W.G. (1906) *Folkways*, New Haven, Yale University Press.

Tacitus P.C. (1914) *Agricola, Germania, Dialogus* (transl. M. Hutton, Loeb Classical Library) in *Tacitus in Five Volumes*, vol. 1, Cambridge, Harvard University Press, revised and reprinted in 1979, 117-215.

Tesser A., J. Achee (1994) "Aggression, Love, Conformity and Other Social Psychological Catastrophes", in R.R. Vallacher, A. Nowak (eds.) *Dynamical Systems in Social Psychology*, New York, Academic Press, 96-109.

Thane P. (1998) "Family and kinship in modern Britain", paper given at the *Second European Social Science History Conference,* Amsterdam, the Netherlands.

Thomson D. (1984) "The decline of social welfare: Falling State support for the elderly since early Victorian times", *Ageing and Society*, 4, 4, 451-482.

Thomson D. (1991) "The welfare of the elderly in the past: a family or community responsibility", in M. Pelling, R.M Smith (eds.) *Life, Death and the Elderly. Historical Perspectives*, Routledge.

Timberlake E., S.S. Chipingu (1992) "Grandmotherhood: Contemporary meaning among African American middle-class grandmothers", *Social Work*, 37, 3, 216-222.

Tobío-Soler C, J.A. Férnandez-Cordón (1996) "Le rôle des grands-mères dans les familles espagnoles où les fammes travaillent", paper given in the meeting *Entre grands-parents et pétits enfants, la force du lien*, Paris, Ecole des Grand-Parents Europèens.

Tobío-Soler C., J.A. Férnandez-Cordón. (1998) "Monoparentalidad, trabajo y familia", *Revista Internacional de Sociología*, 83, 3, 51-85.

Todd E. (1983) *La troisième planète. Structures familiales et systèmes idéologiques*, Paris, Seuil.

Todd E. 1990. *L'invention de l'Europe*, Paris, Seuil.

Tomassini C., D. Wolf, A. Rosina (2003) "Parental Housing Assistance and Parent-Child Proximity in Italy", *Journal of Marriage and Family*, 65, 700-715.

Trumbach R. (1978) *The Rise of the Egalitarian Family. Aristocratic Kinship and Domestic Relations in Eighteenth-Century England*, New York, Academic Press.

UN-ECE/Economic Commission for Europe (1992) *Fertility and Family Surveys in Countries of the ECE Region*, Questionnaire and Codebook, Geneva.

Urrutikoetxea Lizarraga J. (1992) *"En una mesa y compañía": Caserío y familia campesina en la crisis de la "sociedad tradicional". Irún, 1766-1845*, San Sebastián, Universidad de Deusto.

van de Kaa D. (1988) *The second demographic transition revisited: theories and expectations*, Symposium on population change and European society", 7-10 December, Florence.

van der Woude A.M. (1972) "Variations in the size and structure of the household in the United Provinces of the Netherlands in the seventeenth and eighteenth centuries", in P. Laslett (ed.) 299-318.

van Nimwegen N., H. Moors. (1997) *Population ageing and policy options. Attitudes in EU countries*, paper given at the *XXIII International Population Conference* of IUSSP, Beijing.

Wall R. (1978) "The age at leaving home", *Journal of Family History*, 3, 2, 181-202.

Wall R. (1983a) "The household: demographic and economic change in England, 1650-1970", in R. Wall *et al.* (eds.) 493-512.

Wall R. (1983b) "Does owning real property influence the form of the household? An example from rural West Flanders", in R. Wall *et al.* (eds.) 379-408.

Wall R. (1984) "Residential isolation of the elderly: A comparison over time", *Ageing and Society*, 4, 4, 483-503.

Wall R. (1987) "Leaving home and the process of household formation in pre-industrial England", *Continuity and Change*, 2, 1, 77-102.

Wall R. (1995) "Elderly persons and members of their households in England and Wales from preindustrial times to the present", in D. Kertzer, P. Laslett (eds.) 81-106.

Wall R. (1999) "Beyond the Household: Marriage, Household Formation and the Role of Kin and Neighbours", *International Review of Social History*, 44, 1, 55-76.

Wall R., J. Robin, P. Laslett (1983 eds.) *Family Forms in Historic Europe*, Cambridge, Cambridge University Press.

Weber M. (1904) "Die protestantische Ethic und der Geist des Kapitalismus" *Archiv für Sozialwissenschaft und Sozialpolitik*, XX-XXI.

Weber M. (1930) *The Protestant Ethic and the Spirit of Capitalism*. Translated by T. Parsons. New York and London. Published originally in 1904-1905.

Weiler N.S. (1986) "Family security or social security? The family and the elderly in New York State during the 1920s", *Journal of Family History*, 11, 1, 77-95.

Wilkinson D.Y. (1984) "Afro-American women and their families", *Marriage and Family Review*, 7, 3-4, 125-142.

Young M., P. Willmott (1957) *Family and Kinship in East London*, London, Routledge & Kegan Paul.

European Studies of Population

1. J.-P. Gonnot, N. Keilman and C. Prinz: *Social Security, Household, and Family Dynamics in Ageing Societies.* 1995 ISBN 0-7923-3395-0
2. H. van den Brekel and F. Deven (eds.): *Population and Family in the Low Countries 1994.* Selected Current Issues. 1995 ISBN 0-7923-3396-9
3. R. Cliquet and C. Thienpont: *Population and Development.* A Message from the Cairo Conference. 1995 ISBN 0-7923-3763-8
4. H. van den Brekel and F. Deven (eds.): *Population and Family in the Low Countries 1995.* Selected Current Issues. 1996 ISBN 0-7923-3945-2
5. H.A. de Gans: *Population Forecasting 1895–1945.* The Transition to Modernity. 1999 ISBN 0-7923-5537-7
6. D. van de Kaa, H. Leridon, G. Gesano and M. Okólski (eds.): *European Populations: Unity in Diversity.* 1999 ISBN 0-7923-5838-4 (HB)
7. J. de Beer and L. van Wissen (eds.): *Europe: One Continent, Different Worlds.* Population Scenarios for the 21st Century. 1999 ISBN 0-7923-5840-6 (HB)
8. J. de Beer and F. Deven (eds.): *Diversity in Family Formation.* The 2nd Demographic Transition in Belgium and The Netherlands. 2000 ISBN 0-7923-6461-9
9. E. Tabeau, A. van den Berg Jeths and C. Heathcote (eds.): *Forecasting Mortality in Developed Countries: Insight from a Statistical, Demographic and Epidemiological Perspective.* 2001 ISBN 0-7923-6833-9
10. M. Corijn and E. Klijzing (eds.): *Transitions to Adulthood in Europe.* 2001 ISBN 0-7923-6965-3
11. G. Wunsch, M. Mouchart and J. Duchêne (eds.): *The Life Table.* Modelling Survival and Death. 2002 ISBN 1-4020-0638-1
12. H.-P. Blossfeld and A. Timm (eds.): *Who Marries Whom?* Educational Systems as Marriage Markets in Modern Societies. 2003 ISBN 1-4020-1682-4
13. T. Frejka and J.-P. Sardon: *Childbearing Trends and Prospects in Low-Fertility Countries.* A Cohort Analysis. 2004 ISBN 1-4020-2457-6
14. G. Dalla Zuanna and G.A. Micheli (eds.): *Strong Family and Low Fertility: A Paradox?* New Perspectives in Interpreting Contemporary Family and Reproductive Behaviour. 2004 ISBN 1-4020-2836-9

KLUWER ACADEMIC PUBLISHERS – DORDRECHT / BOSTON / LONDON

MIX
Papier aus verantwortungsvollen Quellen
Paper from responsible sources
FSC® C105338

If you have any concerns about our products,
you can contact us on
ProductSafety@springernature.com

In case Publisher is established outside the EU,
the EU authorized representative is:
Springer Nature Customer Service Center GmbH
Europaplatz 3, 69115 Heidelberg, Germany

Printed by Libri Plureos GmbH
in Hamburg, Germany